Descriptive Handbook of Economic Minerals

Rob Kanen BSc (Hons)

ISBN 1450575447

Table of Contents

Introduction and Terminology

Economic minerals are minerals that are mined for profit or industrial use. The Descriptive Handbook of Economic Minerals is a reference book of all economic minerals. There are two listings for each mineral: 1. Physical properties, listing all physical properties and 2. Optical properties, listing all optical properties visible with a polarizing microscope. A tabular listing of commodities and their ores is included.

The following terms are used in the mineral descriptions:

Physical Properties

Commodity – The mineral commodity

Aluminum

Feldspar

Garnet

Lithium

Uranium

Copper

Lead

Zinc

Nickel

Iron

Phosphate

Gold

Silver

Etc.

Formula – The chemical formula i.e. CuS

Crystal System - The crystal system

Isometric or cubic

Hexagonal or trigonal
Tetragonal
Orthorhombic
Monoclinic
Triclinic

Specific Gravity - The relative weight of a mineral as compared to water

Hardness - Hardness of a mineral according to Moh's Hardness Scale
1 - Soft, i.e. Talc
2 - Copper
3 - Calcite
4 - Fluorite
5 - Medium i.e. Apatite
6 - Feldspar
7 - Glass or Quartz
8 - Topaz
9 - Corundum
10 - Hardest natural occurring mineral, Diamond

Cleavage - Cleavage outlines are often distinctive of a mineral.
Cubic Cleavage - Galena
Cleavage/Foliation Masses/Aggregates - Aggregates of planar minerals, such as chlorite and mica in one direction, parallel to cleavage.

Streak - Color of a minerals powder when crushed. Determined by rubbing a mineral on a porcelain plate.
White - most common, feldspar, calcite, garnet
Black - magnetite, uraninite
Green - hornblende, vesuvianite

Color - Natural color of a mineral
Black - Mica, Uraninite, Hematite

White - Plagioclase

Pink - Orthoclase

Green - Chlorite, olivine

Opacity - The transparency of a mineral

Opaque - Does not transmit light

Translucent - Partially transmits light

Transparent - Fully transmits light

Luster - Appearance of a minerals surface

Pearly - Smooth, shiny, white surface, talc, calcite

Vitreous - Glass like, most transparent minerals are vitreous, quartz, garnet

Dull - Does not shine, talc, kaolinite

Adamantine - Brilliant shining surface, diamond

Soapy - Soapy feel, talc, chlorite

Greasy - Greasy feel, topaz, olivine

Silky - Like silk, antigorite, anthophyllite

Waxy - Wax like surface, quartz, serpentine,

Satin - Satin like surface, kaolinite

Submetallic - Shiny, opaque, metallic like surface, mica

Metallic - Shiny, metallic surface, pyrite

Splintery - In splinters, chlorite

Habit - A minerals external form

Cubic - In cubes, pyrite, halite garnet

Polyhedrons/Octohedrons - Garnet, fluorite

Rhombohedral - Rhomb shaped, calcite, chiastolite

Prismatic - Rectangular, square outlines, very common, hornblende, epidote, augite, zircon, sphene

Hexagonal - Hexagonal or triangular outlines, tourmaline, beryl, topaz

Tabular - Thick, flat prismatic outlines, feldspar, biotite, chlorite, epidote, olivine

Columnar - Thin columns, often in aggregates, tourmaline, actinolite/tremolite, hornblende, diopside

Flakes/Plates/Scales - Mostly the fine, microcrystalline varieties. Mica, clay minerals, chlorite, sericite

Fibrous - In fibers, actinolite/tremolite, talc, serpentine, nephrite

Acicular - Fine, needle like crystals, tourmaline, rutile, mimetite (apatite)

Radiating - Outwardly radiating from a center, zeolite, tourmaline, gypsum

Spherulitic - A radiating acicular mass of crystals forming a circular shaped pattern. Common pattern in devitrified volcanic glass

Short - Short

Long - Long

<u>Uses</u> - The main uses and applications for this commodity

Lithium - Greases, ceramics, production of aluminum

Garnet - Abrasives

Feldspar - Manufacture of porcelain

Lead - Pipes, batteries, radiation shielding

Copper - Alloyed for bronzes, brasses, electrical

Nickel - Steel making

<u>Model</u> - The mode of occurrence and/or formation of an economic mineral

Veins - hydrothermal veins, quartz, calcite veins

Pegmatites - V.coarse grained veins of granites, feldspars, tourmaline, micas

Vesicular/Amygdaloidal - Cavities in volcanic rocks

Evaporites - Within sedimentary evaporite basins

Metamorphic (Contact) - Along contact zones of recrystallized rocks, hornfels, skarn

Metamorphic (Regional) - Metamorphic rocks occurring over a large area, gneiss

Igneous - Generally some form of magmatic intrusion, granite, gabbro

Sedimentary - Derived from pre-existing rocks and often laid down in layers, sandstone, shale

Sedimentary Limestones - Deposited in ocean basins due reef building or accumulation of carbonate grains (detrital or precipitation), limestone

Replacement - Hydrothermal replacement deposits

Optical Properties (Reflected Light)

Most of these properties are only discernable with a polarizing microscope, as used by geologists. The opaque ore minerals are examined with reflected light, only properties for reflected light are listed. These are:

Anisotropy - The interference color observable under crossed polars.

Pleochroism - The variation in color as a mineral is rotated under plane polarized light.

Reflection - The reflection under plane polarized light at a specific wavelength based the COM Data File (Henry, 1977)

Color Index - The quantitative color index using the C.I.E. system.

Vickers Hardness Number (VHN) - Hardness of a polished mineral section, with indentation - fractured, slightly fractured, perfect

Polishing Hardness - Another hardness measurement, either arbitrary number system or relative to other minerals

Internal Reflections - The internal display of colors

Extinction - The type of extinction when a mineral is rotated under crossed polarized light.
Parallel - Parallel to cleavage or crystal outlines
Oblique - At an angle to cleavage or crystal outlines
Symmetrical - Symmetrical to crystal shape or cleavage
Straight - Straight extinction
Wavy/Undulose - Undulose extinction

Cleavage - Cleavage outlines are often distinctive of a mineral.
Cubic Cleavage - Galena

Cleavage/Foliation Masses/Aggregates - Aggregates of planar minerals, such as chlorite and mica in one direction, parallel to cleavage.

Twinning - Twinned crystals are often observed in thin section and are distinctive of certain minerals

Polysynthetic Twinning - Complex twinning along multiple lamellae

Simple Twinning - Two crystals sharing a common twin plane

Form - The habit and shape of the mineral grains/crystals

Alteration - Any alteration of the mineral

Associated Minerals - The usual associated minerals

Model - The geological setting or type of mineral deposit

Distinguishing Features - The most characteristic features of a mineral

In addition, optical properties are listed for those minerals which are transparent – viewed with transmitted light, viewed with a transmitted light polarizing microscope.

Optical Properties (Transmitted Light)

Most of these properties are only discernable with a polarizing microscope, as used by geologists.

Relief - The visibility of a mineral in plane polarized light. Usually compared to adjacent minerals or the cementing material used in the slide, such as balsam.

Birefringence - The interference color observable under crossed polars. Refer to a birefringence table to determine values

2V - The angle of the optical axis in biaxial minerals. Measured in an oriented crystal with a Bertrand Lens under crossed polars and high power.

Optical Sign - Most minerals are either uniaxial, with a single optic axis or biaxial, with two optic axes i.e. quartz, calcite, tourmaline are uniaxial, olivine, augite, hornblende, feldspar are biaxial.

Refractive Index - The degree to which a crystal bends light, as it passes through a crystal. The RI varies according to the optical axes. Uniaxial minerals have two directions of RI: 1. Along the ordinary ray; and 2. Along the extraordinary ray. Biaxial minerals have three directions of RI: 1. Nalpha, 2. Nbeta and 3. Ngamma.

Pleochroism - The variation in color as a mineral is rotated under plane polarized light. I.e. hornblende, biotite are strongly pleochroic

Extinction Angle - The type of extinction when a mineral is rotated under crossed polarized light.
Parallel - Parallel to cleavage or crystal outlines
Oblique - At an angle to cleavage or crystal outlines
Symmetrical - Symmetrical to crystal shape or cleavage, i.e. hornblende, hypersthene, dolomite

Cleavage - Cleavage outlines in thin section are often distinctive of a mineral. There may be some difference between observable cleavages in optical properties and physical properties due to recognition of imperfect cleavages or partings.

Amphibole Cleavage - Two at 56" and 124" in cross section
Pyroxene Cleavage - Two at 83" and 97" in cross section
Mica Cleavage - Perfect in one direction
Chlorite Cleavage - Perfect in one direction
Cubic Cleavage - Halite, galena
Cleavage/Foliation Masses/Aggregates - Aggregates of planar minerals, such as chlorite and mica in one direction, parallel to cleavage.

<u>Twinning</u> - Twinned crystals are often observed in thin section and are distinctive of certain minerals

Polysynthetic Twinning Albite Law - Plagioclase feldspars almost always exhibit this. Used to determine Albite/Anorthite content

Simple Twinning, Amphibole - Hornblende often is twinned with two crystals sharing a common twin plane

Simple Twinning, Carlsbad - Very common in K-feldspars

Penetration Twinning - Where two crystals penetrate each other through the center, andalusite, staurolite, cordierite

Mineral Descriptions

Physical Properties
Name: Alabandite
Commodity: Manganese (Mn)
Formula: MnS
Crystal System: Isometric
Color: Black, Lead gray, Brownish gray
Opacity: Opaque
Luster: Sub metallic
Streak: Dark green
SGLow: 3.95
SGHigh: 4.04
HardnessLow: 3.5
HardnessHigh: 4
Cleavage: 3
Direction: [100] perfect, [010] perfect, [001] perfect
Habit: Massive, granular, disseminated
Fracture: Uneven
Other: Important iron alloy for steel making
Comments: Occurs in epithermal sulfide vein deposits

Optical Properties
Name: Alabandite
Commodity: Manganese
Formula: MnS
Crystal System: Cubic
Color: Grey
Pleochroism: None
Anisotropism: Isotropic
Internal Reflections: Common, dark green to brown
Reflectance 546nm: 22.8
Reflectance at 589nm: 22.3
Quant. Color Coordinates (QC): QC=0.301, 00.305, 22.8
Vickers Hardness Number (100g): 240-251, perfect
Polishing Hardness: Equal to sphalerite
Cleavage:
Form: Euhedral cubic crystal and anhedral aggregates
Alteration:
Associated Minerals: Associated with pyrite, chalcopyrite, pyrrhotite, pyrolusite, MN-sphalerite, Mn-carbonate
Distinguishing Features: Cubic crystals, isotropism, internal reflections

Physical Properties

Name: Albite
Commodity: Feldspar
Formula: NaAlSi3O8
Crystal System: Triclinic
Color: White to colorless, bluish, grey, reddish, greenish
Opacity: Transparent to subtranslucent
Luster: Vitreous, pearly
Streak: White
SGLow: 6
SGHigh: 6.5
HardnessLow: 2.6
HardnessHigh: 2.63
Cleavage: 2
Direction: {001} perfect, {010} nearly perfect
Habit: Crystals tabular, small; usually massive, lamellar, or granular; twinning common, simple, multiple, and repeated
Fracture: Conchoidal to uneven; brittle
Other: Manufacture of porcelain, pottery and glass, glazes, paint filler, abrasives
Comments: Manufacture of porcelain, pottery and glass, glazes, paint filler, abrasives

Optical Properties

Name: Albite
Group: Feldspar
Formula: NaAlSi3O8 An0-10%
Crystal System: Triclinic
Color: Colorless
Form: Plates or lathe shaped sections, rarely in phenocrysts
Relief: Low, n considerably <balsam, n=balsam
Birefringence: Weak, 0.009-0.011
2V: 77"-82"
Nalpha or Nord.: 1.527-1.533
NBeta or Nextr.: 1.531-1.537
NGamma: 1.538-1.542
Optical Sign: Biaxial positive
Orientation:
Pleochroism:
Twinning: Polysynthetic twinning according to albite law. Twinning according to carlsbad law either alone or combined. Occasional pericline twinning
Cleavage: Four directions, perfect {001}, distinct {010}, imperfect {110} and {11_0}
Alteration: Alters to clay
Features: RI equal to and less than balsam is distinctive, albite twinning and extinction angles help.
Occurrence: Most abundant in felsic igneous rocks. Widespread occurrence

Physical Properties

Name: Allargentum

Commodity: Silver (Ag)

Formula: Ag1-xSbx(x=0.009-0.16)

Crystal System: Hexagonal

Color: Silver gray

Opacity: Opaque

Luster: Metallic

Streak:

SGLow: 10

SGHigh: 10

HardnessLow: 4

HardnessHigh: 4

Cleavage:

Direction:

Habit:

Fracture:

Other: Used in coinage, jewellery, sterling ware, mirrors, electroplating, batteries and photographic and electronic products

Comments: Occurs in high-grade silver ores

Optical Properties

Name: Allargentum

Commodity: Silver

Formula: Ag1-xSbX

Crystal System: Hexagonal

Color: White, slightly greyish, silver

Pleochroism: None

Anisotropism: Weak

Internal Reflections: None

Reflectance 546nm: Approx. 70

Reflectance at 589nm:

Quant. Color Coordinates (QC):

Vickers Hardness Number (100g):

Polishing Hardness: Greater than silver

Cleavage:

Form: Occurs as lamellar intergrowths in silver

Alteration:

Associated Minerals: Similar to dyscrasite, Ag3Sb

Distinguishing Features: Lamellar intergrowths in silver, especialy from Cobalt, Ontario

Physical Properties

Name: Almandine

Commodity: Garnet

Formula: Fe++3Al2(SiO4)3

Crystal System: Isometric

Color: Deep red, brownish red, brownish black, purplish red

Opacity: Transparent to translucent

Luster: Vitreous to resinous

Streak: White

SGLow: 4.1

SGHigh: 4.3

HardnessLow: 7

HardnessHigh: 7.5

Cleavage: 1

Direction: {110} parting sometimes distinct

Habit: Crystals usually dodecahedrons or trapezohedrons; also in combination or with hexoctahedron; massive; granular

Fracture: Conchoidal to uneven; brittle

Other: Abrasives, gemstones

Comments: Occurs in many types of metamorphic rocks and some igneous rocks and in alluvial and beach deposits

Optical Properties

Name: Almandine

Group: Garnet

Formula: Fe3Al2(SiO4)3

Crystal System: Isometric, cubic

Color: Colorless to pale reddish

Form: Euhedral dodecahedrons in six sided trapezohedrons in eigth sided cross sections. Plygonal grains and aggregates.

Relief: V. high, n>balsam

Birefringence: Isotropic but mat show weak birefringence

2V:

Nalpha or Nord.: 1.778-1.815

NBeta or Nextr.:

NGamma:

Optical Sign:

Orientation:

Pleochroism:

Twinning:

Cleavage: Parting parallel to {110}, irregular fractures

Alteration:

Features: Similar to spinel, which is octohedral. Determination of RI will differentiate garnets.

Occurrence: Predominantly found in metamorphic rocks, especially schists and gneisses

Physical Properties
Name: Alunite
Commodity: Potash
Formula: KAl3(SO4)2(OH)6
Crystal System: Trigonal
Color: White, often discolored greyish, yellowish, reddish, brownish
Opacity: Transparent to nearly opaque
Luster: Vitreous, pearly on base
Streak: White
SGLow: 2.6
SGHigh: 2.7
HardnessLow: 3.5
HardnessHigh: 4
Cleavage: 2
Direction: {0001} distinct, {01-12} in traces
Habit: Crystals rhombohedral, often pseudocubic; as druses or aggregates; usually massive; fibrous
Fracture: Conchoidal; brittle
Other: Used in fertilisers and potassium salts
Comments: Occurs with halite and sylvine in sedimentary basin evaporite deposits

Optical Properties
Name: Alunite
Group: Alunite
Formula: KAl3(OH)6(SO4)2
Crystal System: Hexagonal
Color: Colorless
Form: Fine to coarse aggregates of tabular to pseudo cubic rhombohedral
Relief: Fair, n>balsam
Birefringence: Moderate, 0.020
2V:
Nalpha or Nord.: 1.572
NBeta or Nextr.: 1.592
NGamma:
Optical Sign: Uniaxial positive
Orientation: Length fast
Pleochroism:
Twinning:
Cleavage: Fair in one direction, {0001}
Alteration:
Features: Tabular crystals, birefringence
Occurrence: Hydrothermal alteration product of rhyolites, dacites and andesites.

Physical Properties

Name: Amblygonite

Commodity: Lithium (LI)

Formula: $(Li,Na)Al(PO_4)(F,OH)$

Crystal System: Triclinic

Color: White to greyish white, also colorless, yellowish, pinkish, tan, greenish, bluish

Opacity: Transparent to translucent

Luster: Vitreous to greasy, pearly on cleavage

Streak:

SGLow: 3.08

SGHigh: 3.08

HardnessLow: 5.5

HardnessHigh: 6

Cleavage: 4

Direction: {100} perfect, {110} good, {0-11} distinct, {001} imperfect

Habit: Crystals equant to short prismatic; usually large cleavable masses

Fracture:

Other: Greases, ceramics, production of Aluminum

Comments: Occurs in pegmatites with lepidolite, tourmaline, beryl and other lithium minerals. Rare.

Physical Properties

Name: Amosite (fibrous form of ferro-gedrite) (Grunerite)

Commodity: Asbestos

Formula: $(Mg,Fe)_7Si_8O_{22}(OH)_2$

Crystal System: Orthorhombic

Color: White, grey, greenish, brownish green, clove-brown, dark brown

Opacity: Transparent to nearly opaque

Luster: Vitreous to silky

Streak: Colorless or greyish

SGLow: 2.85

SGHigh: 3.57

HardnessLow: 5.5

HardnessHigh: 6

Cleavage: 3

Direction: {110} perfect, {010} and {100} imperfect

Habit: Fibrous to asbestiform

Fracture:

Other: Heat resistant, used for brake pads and fire retardent materials, roofing and cladding and fire retardent paints

Comments: Fibrous anthophyllite that occurs in metamorphic rocks

Optical Properties

Name: Amosite (Grunerite)

Group: Amphibole

Formula: $(Fe,Mg)_7Si_8O_{22}(OH)_2$

Crystal System: Monoclinic

Color: Neutral

Form: Fibrous to columnar aggregates, sometimes asbestiform. Cross sections are rhombic

Relief: High, n>balsam

Birefringence: Strong, 0.042-0.054. Sections with parallel extinction are first order.

2V: 79"-86"

Nalpha or Nord.: 1.657-1.663

NBeta or Nextr.: 1.684-1.697

NGamma: 1.699-1.717

Optical Sign: Biaxial negative

Orientation: Length slow, r>v weak

Pleochroism:

Twinning: Polysynthetic twinning characteristic {100}

Cleavage: In two directions at 56" and 124".

Extinction: Oblique 10"-15" in longitudinal sections

Alteration:

Features: Series with magnesiocummingtonite and cummingtonite. Smaller extiction angle than cummingtonite. RI higher than tremolite. Anthophyllite has parallel extinction.

Occurrence: Chiefly within metamorphic rocks, particularly banded iron formations and mica schists

Physical Properties

Name: Andalusite

Commodity: Aluminum Silicates

Formula: Al2SiO5=Al[6]Al[5]OSiO4

Crystal System: Orthorhombic

Color: Pink, reddish brown, rose-red, whitish; greyish, yellowish, violet, greenish, blue

Opacity: Transparent to nearly opaque

Luster: Vitreous to sub vitreous

Streak: Colorless

SGLow: 3.13

SGHigh: 3.16

HardnessLow: 6.5

HardnessHigh: 7.5

Cleavage: 2

Direction: {110} distinct, {100} indistinct

Habit: Crystals prismatic, nearly square in cross section; massive, compact

Fracture: Subconchoidal to uneven; brittle

Other: Refractories for steel smelters, glass industry, insulating porcelain

Comments: Widely distributed in schists, gneisses, pegmatites and quartz veins in metamorphic belts and contact aureoles

Optical Properties

Name: Andalusite (Chiastolite)

Group: Sillimanite

Formula: Al2SiO5

Crystal System: Orthorhombic

Color: Colorless, rarely reddish

Form: Euhedral square shaped crystals and columnar aggregates

Relief: High, n>balsam

Birefringence: Weak, 0.007-0.011

2V: 84"

Nalpha or Nord.: 1.629-1.640

NBeta or Nextr.: 1.633-1.644

NGamma: 1.639-1.647

Optical Sign: Biaxial negative

Orientation: Length fast, r>v weak

Pleochroism: Colored varieties are plaochroic from rose red-pale green

Twinning: Penetration twins common

Cleavage: Distinct parallel to {110}. Cross sections show two directions at right angles.

Alteration: Alters to sillimanite

Features: Distinguished from sillimanite by cleavage at right angles, pleochroism, weak birefringence

Occurrence: Widespread within gneisses, schists, slates, hornfelses and other metamorphic rocks

Physical Properties

Name: Andradite

Commodity: Garnet

Formula: Ca3Fe+++2(SiO4)3

Crystal System: Isometric

Color: Shades of yellowish green, green, greenish brown, brown, reddish brown, greyish black

Opacity: Transparent to nearly opaque

Luster: Vitreous to Resinous

Streak:

SGLow: 3.7

SGHigh: 4.1

HardnessLow: 6.5

HardnessHigh: 7

Cleavage: 0

Direction:

Habit: Crystals usually dodecahedrons or trapezohedrons; massive; granular

Fracture: Uneven to conchoidal; brittle

Other: Abrasives, gemstones

Comments: Occurs in many types of metamorphic rocks and some igneous rocks and in alluvial and beach deposits

Optical Properties

Name: Andradite

Group: Garnet

Formula: Ca3Fe2(SiO4)3

Crystal System: Isometric, cubic

Color: Colorless to pale red, pale brown to brown, greenish

Form: Euhedral dodecahedrons in six sided trapezohedrons in eigth sided cross sections. Plygonal grains and aggregates.

Relief: V.high, n>balsam

Birefringence: Isotropic, but may show weak birefringence

2V:

Nalpha or Nord.: 1.857-1.887

NBeta or Nextr.:

NGamma:

Optical Sign:

Orientation:

Pleochroism:

Twinning:

Cleavage: Parting parallel to {110}, irregular fractures

Alteration:

Features: Similar to spinel which is octohedral. Determination of RI will differentiate garnets.

Occurrence: Occurs within contact metamorphic zones

Physical Properties
Name: Anglesite
Commodity: Lead (Pb)
Formula: $PbSO_4$
Crystal System: Orthorhombic
Color: Colorless, white, yellowish, grey, pale shades of green and blue
Opacity: Transparent to opaque
Luster: Adamantine, vitreous or resinous
Streak: Colorless
SGLow: 6.38
SGHigh: 6.38
HardnessLow: 2.5
HardnessHigh: 3
Cleavage: 3
Direction: {001} good, {210} distinct, {010} indistinct
Habit: Crystals varied in habit, thin to thick tabular, prismatic or equant; massive, granular, nodular, stalactitic
Fracture: Conchoidal; brittle
Other: Pipes, alloys, batteries, radiation shielding, pigments
Comments: Associated with galena, cerrusite, smithsonite and sphalerite in oxidized zone of lead deposits

Optical Properties
Name: Anglesite
Group: Barite
Formula: $PbSO_4$
Crystal System: Orthorhombic
Color: Blue, colorless, green, gray, yellow
Form: Anhedral to subhedral crystals
Relief:
Birefringence: Moderate, 0.017
2V: 68"-75.4"
Nalpha or Nord.: 1.878
NBeta or Nextr.: 1.883
NGamma: 1.895
Optical Sign: Biaxial positive
Orientation:
Pleochroism:
Twinning:
Cleavage: Two {001} good, {210} distinct
Extinction:
Alteration:
Features:
Occurrence: Secondary, weathered deposits of lead ore

Physical Properties

Name: Anhydrite

Commodity: Anhydrite (CaSo4)

Formula: CaSO4

Crystal System: Orthorhombic

Color: Colorless, white, grey, bluish, violet, pinkish, reddish, brownish

Opacity: Transparent to translucent

Luster: Vitreous to greasy or pearly

Streak: White or greyish white

SGLow: 2.98

SGHigh: 2.98

HardnessLow: 3.5

HardnessHigh: 3.5

Cleavage: 3

Direction: {010} perfect, {100} nearly perfect, {001} good to imperfect

Habit: Crystals equant, thick tabular or prismatic, rare; usually massive, granular

Fracture: Splintery to uneven; brittle

Other: Used as a fertilizer and manufacture of cements, plasters, sulphates and sulphuric acid

Comments: Occurs in arid climates where it forms deposits in evaporite lakes

Optical Properties

Name: Anhydrite

Group: Evaporites

Formula: CaSO4

Crystal System: Orthorhombic

Color: Colorless

Form: Anhedral to subhedral aggregates of fine to medium grains, occasionally elongate

Relief: Moderate, n>balsam

Birefringence: Strong, 0.044

2V: 42"

Nalpha or Nord.: 1.570

NBeta or Nextr.: 1.576

NGamma: 1.614

Optical Sign: Biaxial positive

Orientation:

Pleochroism:

Twinning: Polysynthetic with {101} Forms angles 42" and 48" to cleavage traces. Sometimes two sets of twin lamellae intersection at 83.5" and 96.5"

Cleavage: Three directions at right angles. Parallel to {100}, {010} and {001}. May show parting parallel to {101) due to twinning

Alteration: Often altered to gypsum

Features: Strong birefringence, rectangular pseudo cubic cleavage distinguishes from gypsum

Occurrence: Found within sedimentary evaporite deposits. Main constituent of anhydrite rock

Physical Properties

Name: Anorthite

Commodity: Feldspar

Formula: Ca2Al2Si2O8

Crystal System: Triclinic

Color: Colorless, white, greyish, reddish

Opacity: Transparent to translucent

Luster: Vitreous

Streak: White

SGLow: 2.74

SGHigh: 2.76

HardnessLow: 6

HardnessHigh: 6.5

Cleavage: 3

Direction: {001} perfect, {010} nearly perfect, {110} imperfect

Habit: Crystals usually short prismatic; massive, cleavable

Fracture: Conchoidal to uneven; brittle

Other: Manufacture of porcelain, pottery and glass. Used for glazes in pottery and as mild abrasive

Comments: Important rock forming mineral widely distributed in igneous and metamorphic rocks

Optical Properties

Name: Anorthite

Group: Feldspar

Formula: CaAl2Si2O8 An90-100%

Crystal System: Triclinic

Color: Colorless

Form: Anhedral to subhedral plates and lathes

Relief: Fair, n>balsam

Birefringence: Weak, 0.012-0.013

2V: 77"-79"

Nalpha or Nord.: 1.573-1.577

NBeta or Nextr.: 1.579-1.585

NGamma: 1.585-1.590

Optical Sign: Biaxial negative

Orientation: r>v

Pleochroism:

Twinning: According to albite, carlsbad and pericline laws, as in albite

Cleavage: Four, perfect {001}, distinct {010}, imperfect {110} and {11_0}

Alteration: Alters to clay

Features: Maximum extinction angles in albite twins and RI are distinctive

Occurrence: Quite rare compared to other plagioclase feldspars. occurs in some lavas and contact metamorphic rocks.

Physical Properties
Name: Antimony
Commodity: Antimony (Sb)
Formula: Sb
Crystal System: Trigonal
Color: Light gray, tin white
Opacity: Opaque
Luster: Metallic
Streak: Lead gray
SGLow: 6.61
SGHigh: 6.72
HardnessLow: 3
HardnessHigh: 3.5
Cleavage: 1
Direction: {0001} perfect
Habit: Massive, lamellar, foliated, network fibrous and columnar aggregates
Fracture: Brittle
Other: Flame resistant properties used in textiles and other materials. Alloying with other metals, particularly lead
Comments: Hydrothermal veins

Optical Properties
Name: Antimony
Commodity: Antimony
Formula: Sb
Crystal System: Trigonal, hexagonal
Color: White
Pleochroism: Weak
Anisotropism: Distinct, yellowish-grey to brownish to bluish-grey
Internal Reflections: None
Reflectance 546nm: 71.1-73.0
Reflectance at 589nm: 70.0-72.1
Quant. Color Coordinates (QC):
Vickers Hardness Number (100g): 84-98
Polishing Hardness: Greater than stibnite, less than arsenic
Cleavage: Often visible
Form: Fine to coarse grained aggregates. Polysynthetic twinning and cleavage often visible.
Alteration:
Associated Minerals: Associated with stibnite, pyrite, arsenopyrite, Co-Ni arsenides and stibarsen as graphic and myrmetitic intergrowths known as allemontite
Distinguishing Features: White color and yellowish-bluish grey anisotropism

Physical Properties

Name: Apatite (Group Name)

Commodity: Phosphates

Formula: Ca5(PO4)3F

Crystal System: Hexagonal

Color: Colorless, white, grey, yellow to yellowish green, green, blue, violet, red, brown

Opacity: Transparent to opaque

Luster: Vitreous to subresinous, silky

Streak: White

SGLow: 3.1

SGHigh: 3.2

HardnessLow: 5

HardnessHigh: 5

Cleavage: 2

Direction: {0001} indistinct, {10-10} trace

Habit: Crystals short to long prismatic, thin to thick tabular or complex; massive compact to coarse granular; fibrous

Fracture: Conchoidal to uneven; brittle

Other: Fertilisers and in manufacture of phosphorous chemicals

Comments: Accessory mineral in igneous and metamorphic rocks, particluarly, pegmatites, metamorphosed limestones and skarns

Optical Properties

Name: Apatite

Group: Apatite

Formula: Ca5(PO4)3F

Crystal System: Hexagonal

Color: Colorless

Form: Minute, hexagonal crystals

Relief: Moderate, n>balsam

Birefringence: Weak, 0.003-0.004

2V:

Nalpha or Nord.: 1.630-1.651

NBeta or Nextr.: 1.633-1.655

NGamma:

Optical Sign: Uniaxial negative

Orientation: Length fast, tabular crystals are length slow

Pleochroism:

Twinning:

Cleavage: Two, imperfect basal {0001}

Alteration:

Features: Hexagonal crystals and birefringence

Occurrence: Widespread accessory mineral in igneous rocks.. In pegmatites, veins, metamorphic limestones and sedimentary beds.

Physical Properties
Name: Argentite
Commodity: Silver (Ag)
Formula: Ag2S
Crystal System: Cubic
Color: Lead-grey to blackish lead-grey
Opacity: Opaque
Luster: Metallic
Streak:
SGLow: 7.2
SGHigh: 7.34
HardnessLow: 2
HardnessHigh: 2.5
Cleavage: 2
Direction: {001} and {011} poor
Habit: Crystals usually cubic or octahedral; dodecahedral; modified and distorted; massive, branching, reticulated; granular
Fracture: Subconchoidal; sectile
Other: Used in coinage, jewellery, sterling ware, mirrors, electroplating, batteries and photographic and electronic products
Comments: Associated with lead, zinc and copper ores. Recovered as a byproduct during smelting of these ores

Optical Properties
Name: Acanthite (Argentite)
Commodity: Silver (Ag)
Formula: Ag2S
Crystal System: Monoclinic
Color: Grey, greenish tint
Pleochroism: Very weak
Anisotropism: Distinct
Internal Reflections: None
Reflectance 546nm: 30.3-31.3
Reflectance at 589nm: 29.0-29.8
Quant. Color Coordinates (QC):
Vickers Hardness Number (100g): 23-26, perfect
Polishing Hardness: Less than most minerals
Cleavage:
Form: Euhedral cubic crystal pseudomorphs after argentite
Alteration: Argentite alters to Acanthite upon cooling at 176"C
Associated Minerals: Associated with galena (often as inclusions), pyrite, sphalerite, tetrahedrite, covellite, proustite, pyragyrite, polybasite
Distinguishing Features: Cubic pseudomorphs after argentite, color

Physical Properties

Name: Argyrodite

Commodity: Germanium

Formula: Ag8GeS6

Crystal System: Orthorhombic ps. cub.

Color: Steel-grey with reddish tint, tarnishes black with purple or bluish tint

Opacity: Opaque

Luster: Metallic

Streak: greyish black

SGLow: 6.20

SGHigh: 6.20

HardnessLow: 2.5

HardnessHigh: 2.5

Cleavage:

Direction:

Habit: Crystals octahedral, dodecahedral; crystal aggregates, botryoidal crusts, massive

Fracture: Conchoidal to uneven; brittle

Other: Electronics

Comments: Occurs with sphalerite, siderite and marcasite. A byproduct collected from cadmium fume during sintering of zinc concentrates

Optical Properties

Name: Argyrodite

Commodity: Germanium (Ge)

Formula: Ag8GeS6

Crystal System: Orthorhombic

Color: Gray-white with a violet tint

Pleochroism: Very weak

Anisotropism: Weak

Internal Reflections:

Reflectance 546nm:

Reflectance at 589nm:

Quant. Color Coordinates (QC):

Vickers Hardness Number (100g):

Polishing Hardness:

Cleavage: None

Form: Crystals octahedral, dodecahedral; crystal aggregates, botryoidal crusts, massive

Alteration:

Associated Minerals:

Distinguishing Features:

Model: In low-temperature polymetallic deposits with silver sulfosalts and in high-temperature Sn-Ag deposits. A byproduct collected from cadmium fume during sintering of zinc concentrates

Physical Properties

Name: Arsenic

Commodity: Arsenic (As)

Formula: As

Crystal System: Trigonal

Color: Lead gray, gray, white

Opacity: Opaque

Luster: Metallic

Streak: Black

SGLow: 5.7

SGHigh: 5.7

HardnessLow: 3.5

HardnessHigh: 3.5

Cleavage: 1

Direction: {0001} perfect

Habit: Lamellar, reniform, nodular

Fracture: Uneven

Other: Used in insecticides and as alloy with copper and lead

Comments: Associated with many ore minerals, particularly in hydrothermal vein deposits

Optical Properties

Name: Arsenic

Commodity: Arsenic (As)

Formula: As

Crystal System: Trigonal, hexagonal

Color: White, tarnishes rapidly

Pleochroism: Weak in air, greyish white to yellow or bluish grey in oil

Anisotropism: Distinct, grey to yellowish grey

Internal Reflections: None

Reflectance 546nm:

Reflectance at 589nm:

Quant. Color Coordinates (QC):

Vickers Hardness Number (100g):

Polishing Hardness: Greater than bismuth and silver

Cleavage: Basal cleavage often visible

Form: Anhedral fine to coarse grained aggregates, twinning and basal cleavage often visible

Alteration:

Associated Minerals: Associated with rammelsburgite, skutterudite, proustite, arsenopyrite. Fine graphic and myrmekitik intergrowths within stibarsen. These are known as allemontite.

Distinguishing Features: The rapid tarnishing, within a few hours, is diagnostic.

Physical Properties

Name: Arsenopyrite

Commodity: Arsenic (As)

Formula: FeAsS

Crystal System: Monoclinic ps. orth.

Color: Silver-white to steel-grey

Opacity: Opaque

Luster: Metallic

Streak: Black

SGLow: 6.07

SGHigh: 6.07

HardnessLow: 5.5

HardnessHigh: 6

Cleavage: 2

Direction: {101} distinct, {010} traces

Habit: Crystals short prismatic, may be elongated parallel to c-axis or b-axis; massive, compact, granular or columnar

Fracture: Uneven; brittle

Other: Used in insecticides and as alloy with copper and lead

Comments: Main ore of arsenic, usually regarded as a gangue mineral. Occurs in epithermal to mesothermal quartz vein deposits with tin, tungsten, gold, silver, sphalerite and pyrite

Optical Properties

Name: Arsenopyrite

Commodity: Arsenic (As)

Formula: FeAsS

Crystal System: Monoclinic ps. orth.

Color: White

Pleochroism: Weak

Anisotropism: Strong, blue to green

Internal Reflections: None

Reflectance 546nm: 51.85-52.2

Reflectance at 589nm: 51.8-53.2

Quant. Color Coordinates (QC): 0.315, 0.320, 52.5

Vickers Hardness Number (100g): 715-1354, slightly fractured

Polishing Hardness: Greater than skutterudite and magnetite, less than pyrite, cobaltite

Cleavage: {101} distinct, {010} traces

Form: Euhedral to subhedral rhomb shaped crystals and anhedral granular masses. Lamellar twinning common.

Alteration:

Associated Minerals: Occurs with pyrite, loellingite, glaucodot, pyrrhotite, chalcopyrite, sphalerite, galena, cobaltite, gold and molybdenite

Distinguishing Features: White color, anisotriism and rhombic crystal form are diagnostic

Model: Main ore of arsenic, usually regarded as a gangue mineral. Occurs in epithermal to mesothermal quartz vein deposits with tin, tungsten, gold, silver, sphalerite and pyrite

Physical Properties
Name: Autunite
Commodity: Uranium (U) and Vanadium (V)
Formula: Ca(UO2)2(PO4)2+10-12H2O
Crystal System: Tetragonal
Color: Sulfur-yellow, lemon-yellow, greenish-yellow, pale green to dark green
Opacity: Transparent to translucent
Luster: Vitreous, pearly on {001}, masses dull
Streak: Pale yellow
SGLow: 3.05
SGHigh: 3.20
HardnessLow: 2
HardnessHigh: 2.5
Cleavage: 2
Direction: {001} perfect; {100} indistinct
Habit: Crystals thin to thick tabular with rectangular or octagonal shape; fan-like groups, crusts, foliated to scaly masses
Fracture: Flexible in thin plates
Other: Radioactve, used as nuclear fuel
Comments: Occurs in the oxidized zone of uranium deposits with torbenite and carnotite

Optical Properties
Name: Autunite
Commodity: Uranium (U) and Vanadium (V)
Formula: Ca(UO2)2(PO4)2+10-12H2O
Crystal System: Tetragonal
Color: Sulfur-yellow, lemon-yellow, greenish-yellow, pale green to dark green
Pleochroism: Transparent to translucent
Anisotropism: Vitreous, pearly on {001}, masses dull
Internal Reflections: Pale yellow
Reflectance 546nm: 2
Reflectance at 589nm: 2.5
Quant. Color Coordinates (QC): 3.05
Vickers Hardness Number (100g): 3.20
Polishing Hardness: 2
Cleavage: {001} perfect; {100} indistinct
Form: Crystals thin to thick tabular with rectangular or octagonal shape; fan-like groups, crusts, foliated to scaly masses
Alteration:
Associated Minerals:
Distinguishing Features: Crystals thin to thick tabular with rectangular or octagonal shape; fan-like groups, crusts, foliated to scaly masses
Model: Occurs in the oxidized zone of uranium deposits with torbenite and carnotite

Physical Properties

Name: Azurite
Commodity: Copper (Cu)
Formula: Cu3(CO3)2(OH)2
Crystal System: Monoclinic
Color: Light blue to very dark blue, usually azure blue
Opacity: Transparent to nearly opaque
Luster: Vitreous
Streak: Blue, lighter than color
SGLow: 3.77
SGHigh: 3.77
HardnessLow: 3.5
HardnessHigh: 4
Cleavage: 3
Direction: {011} slightly imperfect; {100} fair; {110} in traces
Habit: Crystals varied in habit and often modified; tabular or short prismatic, equant or rhombohedral; massive, stalactitic, earthy, nodular concretions
Fracture: Conchoidal; brittle
Other: Alloyed with various metals to produce bronzes and brasses. Used in electrical industries where high electrical and thermal conductivity is required.
Comments: Chalcopyrite is the main ore of copper. Azurite is found in the oxidized zone of copper deposits

Optical Properties

Name: Azurite
Commodity: Copper (Cu)
Formula: Cu3(CO3)2(OH)2
Crystal System: Monoclinic
Color: Grey, masked by internal reflections
Pleochroism: Masked by internal reflections
Anisotropism: Masked by internal reflections
Internal Reflections: Strong blue internal reflections
Reflectance 546nm:
Reflectance at 589nm:
Quant. Color Coordinates (QC):
Vickers Hardness Number (100g):
Polishing Hardness:
Cleavage: {011} slightly imperfect; {100} fair; {110} in traces
Form: Usually platy or anhedral, concretions and crusts
Alteration:
Associated Minerals: Malachite
Distinguishing Features: Color and internal reflections distinctive
Model: Chalcopyrite is the main ore of copper. Azurite is found in the oxidized zone of copper deposits

Physical Properties

Name: Baddeleyite

Commodity: Baddeleyite

Formula: ZrO_2

Crystal System: Monoclinic

Color: Colorless to yellow, green, reddish or greenish brown, brown to nearly black

Opacity:

Luster: Greasy to vitreous

Streak: White to brownish white

SGLow: 5.739

SGHigh: 5.739

HardnessLow: 6.5

HardnessHigh: 6.5

Cleavage: 3

Direction: {001} nearly perfect, {010} and {110} imperfect

Habit: Crystals short to long prismatic; tabular, equant; botryoidal masses

Fracture: Subconchoidal to uneven; brittle

Other: Ceramics, polishing powders, manufacture of zirconium chemicals, refractories

Comments: Found in alluvial and heavy mineral strandline beach deposits with zircon, ilmenite, rutile etc.

Optical Properties

Name: Baddeleyite

Group: Uraninite

Formula: ZrO_2

Crystal System: Monoclinic

Color: Brown, brownish black, colorless, green, greenish brown

Form: Tabular

Relief:

Birefringence: Very high, 0.07

2V: 42"-30"

Nalpha or Nord.: 2.13

NBeta or Nextr.: 2.19

NGamma: 2.2

Optical Sign: Biaxial negative

Orientation: r>v, strong

Pleochroism:

Twinning:

Cleavage: One {001} distinct

Extinction:

Alteration:

Features: Occurs as well-formed coarse sized crystals

Occurrence: Occurs in alluvial gravels and high temperature veins

Physical Properties

Name: Baryte

Commodity: Barytes

Formula: BaSO4

Crystal System: Orthorhombic

Color: Colorless, white to greyish, yellowish to brown, bluish, greenish, reddish

Opacity: Transparent to subtranslucent

Luster: Vitreous to resinous

Streak:

SGLow: 4.50

SGHigh: 4.50

HardnessLow: 3

HardnessHigh: 3.5

Cleavage: 3

Direction: {001} perfect, {210} distinct, {010} imperfect

Habit: Crystals thin to thick tabular; short to long prismatic equant; aggregates, stalactitic, columnar, fibrous, earthy

Fracture: Uneven; brittle

Other: Used in oil and gas industry for drilling muds, manufacture of barium chemicals and as filler and extender in paint and rubber

Comments: Occurs in veins in zinc, copper and lead ore deposits and in sedimentary basins and limestone replacement deposits

Optical Properties

Name: Baryte (Barite)

Group: Barite

Formula: BaSO4

Crystal System: Orthorhombic

Color: Colorless

Form: Granular aggregates

Relief: Fairly high, n>balsam

Birefringence: Weak, 0.012

2V: 36"-37.5"

Nalpha or Nord.: 1.636

NBeta or Nextr.: 1.637

NGamma: 1.648

Optical Sign: Biaxial positive

Orientation:

Pleochroism:

Twinning: Polysynthetic with {110} sometimes

Cleavage: Three directions, parallel to {001} and {110} and at 90" and 78"

Alteration:

Features: Sometimes fluoresces or phosphorescent. Similar to celestite, chemical tests may be necessary.

Occurrence: Widespread vein mineral, with quartz and calcite and in some limestones and sandstones

Physical Properties

Name: Bastnaesite-[Ce]

Commodity: Rare Earths (Ce, La, Y)

Formula: (Ce,La)(CO3)F

Crystal System: Hexagonal

Color: Wax-yellow to reddish brown

Opacity: Transparent to translucent

Luster: Vitreous to greasy

Streak:

SGLow: 4.78

SGHigh: 5.2

HardnessLow: 4

HardnessHigh: 4.5

Cleavage: 2

Direction: {0001} parting, distinct to perfect, {10-10} indistinct

Habit: Tabular or rarely short prismatic; anhedral masses, granular

Fracture: Uneven; brittle

Other: Cerium subgroup is most important of rare earths. Used as a catalyst in petroleum refining, iron-cerium alloys in lighter flints and used in ceramic and glass industries and for manufacture of color televisions

Comments: Occurs in pegmatites and veins and carbonatite plutons

Optical Properties

Name: Bastnaesite-[Ce]

Group: Bastnaesite

Formula: (Ce,La)(CO3)F

Crystal System: Hexagonal

Color: Yellow, reddish-brown

Form: Prismatic, massive, granular

Relief:

Birefringence: Extreme, 0.1

2V:

Nalpha or Nord.: 1.717

NBeta or Nextr.: 1.818

NGamma:

Optical Sign: Uniaxial positive

Orientation:

Pleochroism:

Twinning:

Cleavage: Two, {1011} imperfect, {0001} indistinct

Extinction:

Alteration:

Features:

Occurrence: Contact or alteration zones in alkali rocks

Physical Properties

Name: Bauxite (Group Name)

Commodity: Aluminum (Al)

Formula:

Crystal System:

Color:

Opacity:

Luster:

Streak:

SGLow:

SGHigh:

HardnessLow:

HardnessHigh:

Cleavage:

Direction:

Habit:

Fracture:

Other: Manufacture of Alumininum Alloys

Comments: Mixed with gibbsite, boehmite, iron oxides and silica to form bauxite deposits. Bauxite is pisolitic to massive and forms on the surface from weathering of granitic and gneissic rocks in tropical and subtropical climates

Optical Properties

Name: Bauxite Group (Boehmite)

Group: Bauxite

Formula: ALO(OH)

Crystal System: Orthorhombic

Color: Colorless

Form: Minute, tabular crystals

Relief: Moderate

Birefringence: Moderate, 0.013

2V: Moderate

Nalpha or Nord.: 1.638

NBeta or Nextr.: 1.645

NGamma: 1.651

Optical Sign: Biaxial neg

Orientation:

Pleochroism:

Twinning:

Cleavage: One direction parallel to {010}

Alteration: Dimorphous with diaspore

Features: Birefringence, crystals, closely resembles gibbsite

Occurrence: Occurs as a surface weathering product from tropical weathering in bauxite deposits.

Physical Properties
Name: Berthierite
Commodity: Antimony (Sb)
Formula: $Fe^{2+}Sb_2S_4$
Crystal System: Orthorhombic
Color: Dark steel gray to dark brownish-gray
Opacity: Opaque
Luster: Metallic
Streak: Brownish-gray
SGLow: 4
SGHigh: 4.6
HardnessLow: 2
HardnessHigh: 2.5
Cleavage: 1
Direction: {010} good
Habit: Fibrous, massive, acicular
Fracture: Uneven
Other: Flame resistant properties used in textiles and other materials. Alloying with other metals, particularly lead
Comments: Occurs in low-temperature hydrothermal veins with other antimony minerals

Optical Properties
Name: Berthierite
Commodity: Antimony (Sb)
Formula: $FeSb_2S_4$
Crystal System: Orthorhombic
Color: White-grey with a pink or brown tint
Pleochroism: Strong, brownish pink (a) to greyish white (b) to white (c)
Anisotropism: Very strong, blue to grey-white to brown, pink
Internal Reflections: None
Reflectance 546nm: 36.6-42.0
Reflectance at 589nm: 36.5-41.0
Quant. Color Coordinates (QC):
Vickers Hardness Number (100g): 102-213
Polishing Hardness:
Cleavage:
Form: Euhedral acicular crystals and subhedral aggregates.
Alteration:
Associated Minerals: Asssociated with stibnite, chalcopyrite, pyrite, arsenopyrite, pyrrhotite, gudmundite, sphalerite and galena
Distinguishing Features: Color, pleochroism and anisotropy characteristic

Physical Properties

Name: Berzelianite
Commodity: Selenium (Se)
Formula: Cu2Se
Crystal System: Cubic
Color: Lead-grey with faint bluish tint
Opacity: Opaque
Luster: Metallic
Streak:
SGLow: 6.65
SGHigh: 6.71
HardnessLow: 2
HardnessHigh: 2
Cleavage: 0
Direction:
Habit: As dendritic crusts and thin veinlets
Fracture: Brittle
Other: Used in fade resistant paint pigments, glass production, chemical industry and photo-electric instruments. Alloyed with copper and steel to improve workability
Comments: Selenium minerals are associated with sulphides in metal sulphide ore deposits. Recovered as a byproduct during copper sulphide ore smelting

Optical Properties

Name: Berzelianite
Commodity: Selenium (Se)
Formula: Cu2Se
Crystal System: Cubic
Color: Lead-grey with faint bluish tint
Pleochroism:
Anisotropism:
Internal Reflections:
Reflectance 546nm:
Reflectance at 589nm:
Quant. Color Coordinates (QC):
Vickers Hardness Number (100g):
Polishing Hardness:
Cleavage:
Form:
Alteration:
Associated Minerals:
Distinguishing Features:
Model: Selenium minerals are associated with sulphides in metal sulphide ore deposits. Recovered as a byproduct during copper sulphide ore smelting

Physical Properties

Name: Biotite
Commodity: Mica
Formula: K(Mg,Fe)3(Al,Fe)Si3O10(OH,F)2
Crystal System: Monoclinic
Color: Black, dark brown, reddish brown, green, rarely white
Opacity: Transparent to opaque
Luster: Splendent, submetallic, vitreous
Streak: Colorless
SGLow: 2.7
SGHigh: 3.4
HardnessLow: 2.5
HardnessHigh: 3
Cleavage: 1
Direction: {001} perfect
Habit: Crystals tabular or short prismatic, pseudohexagonal outline; disseminated, massive aggregates
Fracture: Thin laminae flexible to brittle
Other: Used for insulating in electrical equipment, in lubricants, wall finishes, artificial stone
Comments: Widely distributed in igneous rocks, particularly granites and pegmatites and metamorphic schists and gneisses. Also in sedimentary sandstones and mudstones

Optical Properties

Name: Biotite
Group: Mica
Formula: K(Mg,Fe)3(Al,Fe)Si3O10(OH,F)2
Crystal System: Monoclinic
Color: Brown, yellowish-brown, reddish-drown, olive green, green
Form: Euhedral thick tabular six sided crystals and elongated plates
Relief: Fair, n>balsam
Birefringence: Strong, 0.033-0.059
2V: 0"-25"
Nalpha or Nord.: 1.541-1.579
NBeta or Nextr.: 1.574-1.638
NGamma: 1.574-1.638
Optical Sign: Biaxial negative
Orientation: Length slow, r>v or r<v weak
Pleochroism: Strong pleochroism from brown - yellowish brown or reddish brown and olive green - green
Twinning: According to mica law {110}
Cleavage: One perfect {001}
Alteration: Alters to chlorite
Features: Similar to phlogopite but darker and stronger pleochroism
Occurrence: Widespread within felsic igneous rocks, gneisses, schists and low grade metamorphic rocks and contact metamorphic zones

Physical Properties

Name: Bismuth
Commodity: Bismuth (Bi)
Formula: Bi
Crystal System: Trigonal
Color: Silver-white with pinkish tint; tarnishes iridescent
Opacity: Opaque
Luster: Metallic
Streak: Silver-white
SGLow: 9.70
SGHigh: 9.83
HardnessLow: 2
HardnessHigh: 2.5
Cleavage: 3
Direction: {0001} perfect, {10-11} good, {10-14} imperfect
Habit: Crystals indistinct; massive, reticulated, branching, foliated
Fracture: Brittle, sectile
Other: Low melting point alloys for safety equipment. Alloying with lead and tin to improve casting properties
Comments: Minor amounts in veins associated with silver, lead, zinc and tin deposits in metamorphic belts

Optical Properties

Name: Bismuth
Commodity: Bismuth (Bi)
Formula: Bi
Crystal System: Trigonal
Color: Silver-white with pinkish tint; tarnishes iridescent
Pleochroism:
Anisotropism:
Internal Reflections:
Reflectance 546nm:
Reflectance at 589nm:
Quant. Color Coordinates (QC):
Vickers Hardness Number (100g):
Polishing Hardness:
Cleavage:
Form:
Alteration:
Associated Minerals:
Distinguishing Features:
Model: Minor amounts in veins associated with silver, lead, zinc and tin deposits in metamorphic belts

Physical Properties

Name: Bismuthinite

Commodity: Bismuth (Bi)

Formula: Bi2S3

Crystal System: Orthorhombic

Color: Lead-grey to tin-white

Opacity: Opaque

Luster: Metallic

Streak: Lead-grey

SGLow: 6.78

SGHigh: 6.78

HardnessLow: 2

HardnessHigh: 2

Cleavage: 3

Direction: {010} perfect, {100} and {110} imperfect

Habit: Crystals stout to slender prismatic, acicular, sometimes dipyramidal; massive, foliated, fibrous

Fracture: Flexible, sectile

Other: Low melting point alloys for safety equipment. Alloying with lead and tin to improve casting properties

Comments: Associated with magnetite, pyrite, chalcopyrite, galena and sphalerite and tin and tungsten ores in veins in plutonic rocks and metamorphic belts

Optical Properties

Name: Bismuthinite

Commodity: Bismuth (Bi)

Formula: Bi2S3

Crystal System: Orthorhombic

Color: Lead-grey to tin-white

Pleochroism:

Anisotropism:

Internal Reflections:

Reflectance 546nm:

Reflectance at 589nm:

Quant. Color Coordinates (QC):

Vickers Hardness Number (100g):

Polishing Hardness:

Cleavage:

Form:

Alteration:

Associated Minerals:

Distinguishing Features:

Model: Associated with magnetite, pyrite, chalcopyrite, galena and sphalerite and tin and tungsten ores in veins in plutonic rocks and metamorphic belts

Physical Properties

Name: Bismutite
Commodity: Bismuth (Bi)
Formula: Bi2(CO3)O2
Crystal System: Tetragonal
Color: Yellowish white to brownish yellow to brown; greenish, grey to black
Opacity: Transparent to translucent
Luster: Vitreous to pearly to dull
Streak:
SGLow: 6.1
SGHigh: 7.7
HardnessLow: 2.5
HardnessHigh: 3.5
Cleavage: 1
Direction: {001} distinct
Habit: Massive; scaly to lamellar aggregates, fibrous or chalcedonic crusts
Fracture:
Other: Low melting point alloys for safety equipment. Alloying with lead and tin to improve casting properties
Comments: Associated with magnetite, pyrite, chalcopyrite, galena and sphalerite and tin and tungsten ores in veins in plutonic rocks and metamorphic belts

Optical Properties

Name: Bismutite
Commodity: Bismuth (Bi)
Formula: Bi2(CO3)O2
Crystal System: Tetragonal
Color: Yellowish white to brownish yellow to brown; greenish, grey to black
nisotropism:
Internal Reflections:
Reflectance 546nm:
Reflectance at 589nm:
Quant. Color Coordinates (QC):
Vickers Hardness Number (100g):
Polishing Hardness:
Cleavage:
Form:
Alteration:
Associated Minerals:
Distinguishing Features:
Model: Associated with magnetite, pyrite, chalcopyrite, galena and sphalerite and tin and tungsten ores in veins in plutonic rocks and metamorphic belts

Physical Properties

Name: Bixbyite

Commodity: Manganese (Mn)

Formula: $(Mn^{+++},Fe^{+++})_2O_3$

Crystal System: Isometric

Color: Black

Opacity: Opaque

Luster: Metallic

Streak: Black

SGLow: 4.95

SGHigh: 4.95

HardnessLow: 6

HardnessHigh: 6.5

Cleavage: 3

Direction: {111} imperfect, {111} imperfect, {111} imperfect

Habit: Fine cystalline, granular

Fracture: Uneven

Other: Important iron alloy for steel making

Comments: Pneumatolytic, hydrothermal, and metamorphic rocks

Optical Properties

Name: Bixbyite

Commodity: Mn

Formula: $(MnFe)_2O_3$

Crystal System: Cubic

Color: Grey with cream to yellow tint

Pleochroism: Very weak in oil, usually absent

Anisotropism: None, isotropic

Internal Reflections: None

Reflectance 546nm: 22.2

Reflectance at 589nm: 22.0

Quant. Color Coordinates (QC): 0.308, 0.316, 22.1

Vickers Hardness Number (100g): 946-1402, perfect

Polishing Hardness: Greater than hausmanite, equal to braunite

Cleavage: Cleavage {111} common

Form: Euhedral cubic crystals and granular aggregates, lamellar twinning and zonal growth often present

Alteration:

Associated Minerals: Associated with hematite, braunite, pyrolusite, hausmannite

Distinguishing Features: Cubic crystals

Physical Properties

Name: Bohmite (Boehmite)

Commodity: Aluminum (Al)

Formula: AlO(OH)

Crystal System: Orthorhombic

Color:

Opacity:

Luster:

Streak:

SGLow:

SGHigh:

HardnessLow:

HardnessHigh:

Cleavage:

Direction:

Habit:

Fracture:

Other: Manufacture of Alumininum Alloys

Comments: Mixed with gibbsite, boehmite, iron oxides and silica to form bauxite deposits. Bauxite is pisolitic to massive and forms on the surface from weathering of granitic and gneissic rocks in tropical and subtropical climates

Optical Properties

Name: Bohmite (Boehmite)

Commodity: Aluminum (Al)

Formula: AlO(OH)

Crystal System: Orthorhombic

Color:

Pleochroism:

Anisotropism:

Internal Reflections:

Reflectance 546nm:

Reflectance at 589nm:

Quant. Color Coordinates (QC):

Vickers Hardness Number (100g):

Polishing Hardness:

Cleavage:

Form:

Alteration:

Associated Minerals:

Distinguishing Features:

Model: Mixed with gibbsite, boehmite, iron oxides and silica to form bauxite deposits. Bauxite is pisolitic to massive and forms on the surface from weathering of granitic and gneissic rocks in tropical and subtropical climates

Physical Properties

Name: Boracite

Commodity: Borates

Formula: Mg3B7O13Cl

Crystal System: Orthorhombic ps. cub. or ps. tet.

Color: Colorless, white, grey, yellow, pale green to dark green, bluish green

Opacity: Transparent to translucent

Luster: Vitreous

Streak:

SGLow: 2.95

SGHigh: 2.95

HardnessLow: 7

HardnessHigh: 7.5

Cleavage: 0

Direction:

Habit: Crystals cubic, dodecahedral, tetrahedral

Fracture: Conchoidal to uneven

Other: Used in manufacture of insulating fibreglass, as fluxes for manufacture of glass and enamels. Borax is used in cloth manufacture and tanning, soap and glue, as preservatives and antiseptics

Comments: Precipitated in evaporite basins in arid climates. Associated with saline deposits of halite and gypsum

Optical Properties

Name: Boracite

Group: Borates

Formula: Mg3B7O13Cl

Crystal System: Orthorhombic ps. cub. or ps. tet.

Color: Blue green, colorless, gray, yellow

Form: Finely crystalline, disseminated

Relief:

Birefringence: Weak to moderate, 0.01-0.0108

2V:

Nalpha or Nord.: 1.658-1.6622

NBeta or Nextr.: 1.662-1.667

NGamma: 1.668-1.673

Optical Sign: Biaxial positive

Orientation:

Pleochroism:

Twinning:

Cleavage: None

Extinction:

Alteration:

Features:

Occurrence: Associated with evaporites

Physical Properties

Name: Borax

Group: Borax

Formula: $Na_2B_4O_5(OH)_4+8H_2O$

Crystal System: Monoclinic

Color: Colorless, white, greyish, greenish, bluish

Opacity: Transparent to opaque

Luster: Vitreous, sometimes earthy

Streak:

SGLow: 1.715

SGHigh: 1.715

HardnessLow: 2

HardnessHigh: 2.5

Cleavage: 2

Direction: {100} perfect, {110} imperfect

Habit: Crystals short prismatic, somewhat tabular

Fracture: Conchoidal; brittle

Other: Associated with evaporite minerals in evaporite deposits

Comments: Soluble in water

Optical Properties

Name: Borax

Group: Borates

Formula: $Na_2B_4O_5(OH)_4+8H_2O$

Crystal System: Monoclinic

Color: White to greyish, greenish, bluish

Form: Stubby, prismatic crystals

Relief: Low to moderate, n<balsam

Birefringence: Moderate to strong, 0.025

2V: 40"

Nalpha or Nord.: 1.447

NBeta or Nextr.: 1.469

NGamma: 1.472

Optical Sign: Biaxial negative

Orientation: Strong crossed dispersion, r>v, optic plane at right angles to {010}

Pleochroism:

Twinning:

Cleavage: Two {100} perfect, {110} fair, {010} traces

Alteration:

Features: Soluble in water

Occurrence: Associated with evaporite minerals in evaporite deposits

Physical Properties
Name: Bornite
Commodity: Copper (Cu)
Formula: Cu5FeS4
Crystal System: Cubic
Color: Copper-red or bronze, tarnishes iridescent purplish
Opacity: Opaque
Luster: Metallic
Streak: Light greyish black
SGLow: 5.079
SGHigh: 5.079
HardnessLow: 3
HardnessHigh: 3
Cleavage: 1
Direction: {111} in traces
Habit: Crystals cubic, octahedral or dodecahedral, rare; massive, granular
Fracture: Conchoidal to uneven
Other: Alloyed with various metals to produce bronzes and brasses. Used in electrical industries where high electrical and thermal conductivity is required.
Comments: Chalcopyrite is the main ore of copper. In veins with other sulphides and with bornite and pyrite in porphyry copper deposits.

Optical Properties
Name: Bornite
Commodity: Copper (Cu)
Formula: Cu5FeS4
Crystal System: Cubic
Color: Pinkish brown to orange, tarnishes purplish, violet or iridescent
Pleochroism: Weak, may be visible on grain boundaries
Anisotropism: Very weak
Internal Reflections: None
Reflectance 546nm: 21.3
Reflectance at 589nm: 24.5
Quant. Color Coordinates (QC): 0.350, 0.338, 22.0
Vickers Hardness Number (100g): 95-105, slightly fractured
Polishing Hardness: Greater than galena, chalcocite and less than chalcopyrite
Cleavage: Sometimes, {111} in traces
Form: Often as irregular plycrystalline aggregates and euhedral cubic crystals
Alteration: Lamellar exsolution and replacement textures with chalcopyrite, enargite and digenite common. Alters along grain boudaries and fractures to covellite
Associated Minerals: Associated with pyrite, chalcopyrite, enargite, digenite, covellite, linnaeite, sphalerite, chalcopyrite, tetrahedrite and other Pb-Sb sulfosalts
Distinguishing Features: Bronze color and purplish tarnish distinctive
Model: Chalcopyrite is the main ore of copper. In veins with other sulphides and with bornite and pyrite in porphyry copper deposits.

Physical Properties

Name: Boulangerite

Commodity: Antimony (Sb)

Formula: Pb5Sb4S11

Crystal System: Orthorhombic

Color: Lead gray, blue gray, gray

Opacity: Opaque

Luster: Metallic

Streak: Reddish-brown

SGLow: 5.7

SGHigh: 6.3

HardnessLow: 2.5

HardnessHigh: 2.5

Cleavage: 2

Direction: {001} indistinct}, {010} indistinct

Habit: Acicular, massive, fibrous, plumose

Fracture: Brittle

Other: Flame resistant properties used in textiles and other materials. Alloying with other metals, particularly lead

Comments: Low temperature hydrothermal veins

Optical Properties

Name: Boulangerite

Commodity: Pb, Sb

Formula: Pb5Sb4S11

Crystal System: Monoclinic

Color: White with bluish grey tint

Pleochroism: Distinct, grey white to green grey

Anisotropism: Distinct, tan to brown to bluish grey

Internal Reflections: rare, red

Reflectance 546nm: 41.8-37.4

Reflectance at 589nm: 40.7-36.5

Quant. Color Coordinates (QC): 0.303, 0.312, 41.4

Vickers Hardness Number (100g): 92-125, slightly fractured

Polishing Hardness: Less than galena

Cleavage:

Form: Occurs as granular and fibrous aggregates

Alteration:

Associated Minerals: With galena, sphalerite, chalcopyrite, tetrahedrite and other Pb-Sb sulfosalts

Distinguishing Features: Color, pleochroism, anisotropy and fibrous crystals distinctive

Physical Properties

Name: Bournonite

Commodity: Copper (Cu)

Formula: PbCuSbS3

Crystal System: Orthorhombic

Color: Steel-grey to iron-black

Opacity: Opaque

Luster: Metallic

Streak: Steel-grey to iron-black

SGLow: 5.83

SGHigh: 5.83

HardnessLow: 2.5

HardnessHigh: 3

Cleavage: 3

Direction: {010} imperfect, {100} and {001} less perfect

Habit: Crystals short prismatic or tabular; massive, granular

Fracture: Subconchoidal to uneven; brittle

Other: Alloyed with various metals to produce bronzes and brasses. Used in electrical industries where high electrical and thermal conductivity is required.

Comments:

Optical Properties

Name: Bournonite

Commodity: Copper (Cu)

Formula: PbCuSbS3

Crystal System: Orthorhombic

Color: Steel-grey to iron-black

Pleochroism: Opaque

Anisotropism: Metallic

Internal Reflections: Steel-grey to iron-black

Reflectance 546nm: 2.5

Reflectance at 589nm: 3

Quant. Color Coordinates (QC): 5.83

Vickers Hardness Number (100g): 5.83

Polishing Hardness: 3

Cleavage: {010} imperfect, {100} and {001} less perfect

Form: Crystals short prismatic or tabular; massive, granular

Alteration:

Associated Minerals:

Distinguishing Features: Crystals short prismatic or tabular; massive, granular

Physical Properties

Name: Braunite

Commodity: Manganese (Mn)

Formula: Mn7SiO12

Crystal System: Tetragonal

Color: Brownish-black, black, steel-grey

Opacity: Opaque

Luster: Submetallic

Streak:

SGLow: 4.72

SGHigh: 4.83

HardnessLow: 6

HardnessHigh: 6.5

Cleavage: 1

Direction: {112} perfect

Habit: Crystals pyramidal, striated; massive, granular

Fracture: Conchoidal to uneven; brittle

Other: Important iron alloy for steel making

Comments: Occurs in veins with other manganese minerals

Physical Properties

Name: Brucite

Commodity: Magnesium (Mg)

Formula: Mg(OH)2

Crystal System: Trigonal

Color:

Opacity:

Luster:

Streak:

SGLow:

SGHigh:

HardnessLow:

HardnessHigh:

Cleavage:

Direction:

Habit:

Fracture:

Other: Alloyed with aluminum to im,prove strength and corrosion resistance. Manufacture of Iron nodules and desulphuring. Used in petrol, zirconium and titanium production and magnesium die-castings

Comments: Occurs in sedimentary basins in dolomite limestones and in veins with talc and calcite in serpentinite

Optical Properties

Name: Brucite

Group: Bauxite

Formula: Mg(OH)2

Crystal System: Hexagonal

Color: Colorless

Form: Plates or scaley aggregates that appear fibrous

Relief: Fair, n>balsam

Birefringence: Moderate 0.019, anomolous reddish brown hue

2V:

Nalpha or Nord.: 1.566

NBeta or Nextr.: 1.585

NGamma:

Optical Sign: Uniaxioal positive

Orientation: Length fast

Pleochroism:

Twinning:

Cleavage: Perfect one direction, {0001}

Alteration: Often altered to hydro magnesite

Features: Cleavage, anomolous interference colours

Occurrence: Occurs within metamorphic limestones as an alteration of periclase

Physical Properties
Name: Breithauptite
Commodity: Nickel (Ni), Antimony (Sb)
Formula: NiSb
Crystal System: Hexagonal
Color: Copper-red, light brown-red
Opacity: Opaque
Luster: Metallic
Streak: Reddish-brown
SGLow: 8.23
SGHigh: 8.23
HardnessLow: 3.5
HardnessHigh: 4
Cleavage: None
Direction: None
Habit: Tabular, massive, reniform
Fracture: Uneven
Other: Nickel is important Iron alloy for making stainless steel due to corrosion resistence, also non-ferrous alloys. Non toxic, used for food handling and pharmaceutical equipment. Used for electroplating steels and as a base for chromium plate
Comments: Associated with Co-Ni-Ag ores in hydrothermal calcite veins

Optical Properties
Name: Breithauptite
Commodity: Nickel (Ni)
Formula: NiSb
Crystal System: Hexagonal
Color: Pink with violet tint
Pleochroism: Strong pink to pinkish violet
Anisotropism: Very strong, bluish green to bluish Gray to violet red
Internal Reflections: None
Reflectance 546nm: 36.9-48.2
Reflectance at 589nm: 43.7-53.0
Quant. Color Coordinates (QC):
Vickers Hardness Number (100g): 412-584
Polishing Hardness: Less than niccolite, rammelsburgite
Cleavage:
Form: Euhedral and subhedral crystals, often zoned common
Alteration:
Associated Minerals: With niccolite, silver safflorite, galena, chromite, pentlandite, pyrrhotite, Ag sulfosalts
Distinguishing Features: Pink color and anisotropy characteristicc
Model: Associated with Co-Ni-Ag ores in hydrothermal calcite veins

Physical Properties

Name: Calamine (Hemimorphite)

Commodity: Zinc (Zn)

Formula: $Zn_4Si_2O_7(OH)_2+H_2O$

Crystal System: Orthorhombic

Color: White, colorless, pale blue, greenish, grey, yellowish, brown

Opacity: Transparent to translucent

Luster: Vitreous

Streak: Colorless

SGLow: 3.4

SGHigh: 3.5

HardnessLow: 4.5

HardnessHigh: 5

Cleavage: 3

Direction: {110} perfect, {101} imperfect, {001} traces

Habit: Crystals thin tabular, vertically striated, doubly terminated crystals distinctly hemimorphic, often in fan-shaped groups

Fracture: Subconchoidal to uneven; brittle

Other: Alloyed to make brass and zinc die castings. Corrosion resistant coatings on steel (galvanizing) and iron. Manufacture of corrosion resitant paints, pigments and fillers etc.

Comments: Occurs with other zinc ores in veins and replacement deposits

Optical Properties

Name: Calamine (Hemimorphite)

Commodity: Zinc (Zn)

Formula: $Zn_4Si_2O_7(OH)_2+H_2O$

Crystal System: Orthorhombic

Color: White, colorless, pale blue, greenish, grey, yellowish, brown

Pleochroism: Transparent to translucent

Anisotropism: Vitreous

Internal Reflections: Colorless

Reflectance 546nm: 4.5

Reflectance at 589nm: 5

Quant. Color Coordinates (QC): 3.4

Vickers Hardness Number (100g): 3.5

Polishing Hardness: 3

Cleavage: {110} perfect, {101} imperfect, {001} traces

Form: Crystals thin tabular, vertically striated, doubly terminated crystals distinctly hemimorphic, often in fan-shaped groups

Alteration:

Associated Minerals:

Distinguishing Features: Crystals thin tabular, vertically striated, doubly terminated crystals distinctly hemimorphic, often in fan-shaped groups

Model: Occurs with other zinc ores in veins and replacement deposits

Physical Properties

Name: Calaverite

Commodity: Gold (Au) and Tellurium (Te)

Formula: AuTe2

Crystal System: Monoclinic

Color: Brass-yellow to silver-white

Opacity: Opaque

Luster: Metallic

Streak: Greenish to yellowish grey

SGLow: 9.10

SGHigh: 9.40

HardnessLow: 2.5

HardnessHigh: 3

Cleavage: 0

Direction:

Habit: Crystals lath-like or bladed, short prisms striated parallel to elongation; massive, granular

Fracture: Subconchoidal to uneven; brittle

Other: Monetary use, jewellery, electronics and for decoration

Comments: Occurs in hypothermalin quartz veins associated with gold ores. Rare.

Optical Properties

Name: Calaverite

Commodity: Gold (Au) and Tellurium (Te)

Formula: AuTe2

Crystal System: Monoclinic

Color: Brass-yellow to silver-white

Pleochroism: Opaque

Anisotropism: Metallic

Internal Reflections: Greenish to yellowish grey

Reflectance 546nm: 2.5

Reflectance at 589nm: 3

Quant. Color Coordinates (QC): 9.10

Vickers Hardness Number (100g): 9.40

Polishing Hardness: 0

Cleavage:

Form: Crystals lath-like or bladed, short prisms striated parallel to elongation; massive, granular

Alteration:

Associated Minerals:

Distinguishing Features: Crystals lath-like or bladed, short prisms striated parallel to elongation; massive, granular

Model: Occurs in hypothermalin quartz veins associated with gold ores. Rare.

Physical Properties

Name: Calcite

Commodity: Calcium Carbonate (CaCO3)

Formula: CaCO3

Crystal System: Trigonal

Color: Colorless or white, grey, yellow, brown, red, green, blue, black

Opacity: Transparent to translucent

Luster: Vitreous to pearly, dull

Streak: White to greyish

SGLow: 2.710

SGHigh: 2.710

HardnessLow: 3

HardnessHigh: 3

Cleavage: 3

Direction: {10-11} perfect, {01-12} and {0001} parting

Habit: Crystals varied, scalenohedrons and rhombohedrons common; massive, granular, stalactitic, fibrous

Fracture: Conchoidal; brittle

Other: Used in cement and lime, as a smelting flux, in printing processes, as soil dressing. Marble for building and ornamental stones

Comments: Occurs in sedimentary basins where it is a rockforming mineral. In large deposits of limestone, chalk and metamorphosed limestone marble

Optical Properties

Name: Calcite

Group: Calcite

Formula: CaCO3

Crystal System: Hexagonal

Color: Colorless

Form: Anhedral fine to coarse aggragates, oolitic, spherulitic, organic structure

Relief: High in long direction, low in short direction

Birefringence: Extreme, 0.172

2V:

Nalpha or Nord.: 1.486

NBeta or Nextr.: 1.658

NGamma:

Optical Sign: Uniaxial negative

Orientation:

Pleochroism:

Twinning: Polysynthetic twinning on {011_2}

Cleavage: Three rhombohedral {10-11} perfect, {01-12} and {0001} parting

Alteration: Often replaced by quartz

Features: Dolomite, siderite and magnesite are very similar. Dolomite is usually subhedral to euhedral. Siderite usually has iron stainings. Staining may be required to distinguish from magnesite.

Occurrence: The major constituent of sedimentary and metamorphic limestones. The most common vein mineral, with quartz

Physical Properties
Name: Carnallite
Commodity: Magnesium (Mg)
Formula: KMgCl3+6H2O
Crystal System: Orthorhombic
Color: Colorless, white, yellow, blue
Opacity: Transparent to translucent
Luster: Greasy
Streak:
SGLow: 1.602
SGHigh: 1.602
HardnessLow: 2.5
HardnessHigh: 2.5
Cleavage: 0
Direction:
Habit: Crystals thick tabular or pseudohexagonal pyramidal; massive, granular
Fracture: Conchoidal
Other: Alloyed with aluminum to improve strength and corrosion resistance. Manufacture of Iron nodules and desulphuring. Used in petrol, zirconium and titanium production and magnesium die-castings
Comments: Occurs with halite and sylvine in sedimentary basin evaporite deposits

Optical Properties
Name: Carnallite
Group: Evaporite
Formula: KMgCl3·6(H2O)
Crystal System: Orthorhombic
Color: Colorless
Form: Granular, pseudo-hexagonal, fibrous
Relief:
Birefringence: Strong, 0.027
2V: 70"
Nalpha or Nord.: 1.467
NBeta or Nextr.: 1.4755
NGamma: 1.494
Optical Sign: Biaxial positive
Orientation:
Pleochroism:
Twinning:
Cleavage: None
Extinction: None
Alteration:
Features:
Occurrence: Occurs within evaporite basins

Physical Properties

Name: Carnotite

Commodity: Uranium (U)

Formula: $K_2(UO_2)_2V_2O_8+3H_2O$

Crystal System: Monoclinic

Color: Bright yellow to golden yellow, greenish yellow

Opacity:

Luster: Pearly, masses dull or earthy

Streak:

SGLow: 4.70

SGHigh: 4.70

HardnessLow:

HardnessHigh:

Cleavage: 1

Direction: {001} perfect, micaceous

Habit: Crystals microscopic, rhomboidal or diamond-shaped outline; disseminated, coatings, powdery, masses, crusts

Fracture:

Other: Radioactve, used as nuclear fuel

Comments: Secondary mineral in sedimentary rocks and also associated with uraninite in uraninite deposits

Optical Properties

Name: Carnotite

Commodity: Vanadium (V) and Uranium (U)

Formula: $K_2(UO_2)_2V_2O_8+3H_2O$

Crystal System: Monoclinic

Color: Bright yellow to golden yellow, greenish yellow

Pleochroism:

Anisotropism: Pearly, masses dull or earthy

Internal Reflections:

Reflectance 546nm:

Reflectance at 589nm:

Quant. Color Coordinates (QC): 4.70

Vickers Hardness Number (100g): 4.70

Polishing Hardness: 1

Cleavage: {001} perfect, micaceous

Form: Crystals microscopic, rhomboidal or diamond-shaped outline; disseminated, coatings, powdery, masses, crusts

Alteration:

Associated Minerals:

Distinguishing Features: Crystals microscopic, rhomboidal or diamond-shaped outline; disseminated, coatings, powdery, masses, crusts

Model: Secondary mineral in sedimentary rocks and also associated with uraninite in uraninite deposits

Physical Properties
Name: Carrollite
Commodity: Cobalt (Co)
Formula: Cu(Co,Ni)2S4
Crystal System: Cubic
Color: Light grey to steel-grey, rapidly tarnishes to copper-red or violet-grey
Opacity: Opaque
Luster: Metallic
Streak:
SGLow: 4.5
SGHigh: 4.8
HardnessLow: 4.5
HardnessHigh: 5.5
Cleavage: 1
Direction: {001} imperfect
Habit: Crystals octahedral; massive, compact, granular
Fracture: Subconchoidal to uneven
Other: Alloying to produce high temperature steels and magnetic alloys. Used as a catalyst in the chemical industries and as cement in sintered carbide cutting equipment
Comments: Sometimes occurs in small amounts in copper ores. Cobalt is extracted as a byproduc during lead, copper and nickel smelting

Optical Properties
Name: Carrollite
Commodity: Cobalt (Co)
Formula: Cu(Co,Ni)2S4
Crystal System: Cubic
Color: Creamy white, may have slight pinkish tint
Pleochroism: None
Anisotropism: None
Internal Reflections: None
Reflectance 546nm: 45
Reflectance at 589nm:
Quant. Color Coordinates (QC):
Vickers Hardness Number (100g): 525-542
Polishing Hardness: Greater than chalcopyrite, less than pyrite
Cleavage: {001} imperfect
Form: Euhedral to subhedral octohedrons, anhedral granular masses
Alteration:
Associated Minerals: With copper minerals, chalcopyrite, bornite, chalcocite, diggenite, pyrrhotite, siegenite, cobalt-pyrite
Distinguishing Features: Crystals octahedral; massive, compact, granular. Occurrence with copper minerals
Model: Sometimes occurs in small amounts in copper ores. Cobalt is extracted as a byproduc during lead, copper and nickel smelting

Physical Properties

Name: Cassiterite

Commodity: Tin (Sn)

Formula: SnO_2

Crystal System: Tetragonal

Color: Brown to brownish black, black, colorless, grey, yellowish, greenish, red

Opacity: Transparent to nearly opaque

Luster: Adamantine, vitreous

Streak: White, greyish, brown

SGLow: 6.99

SGHigh: 6.99

HardnessLow: 6

HardnessHigh: 7

Cleavage: 4

Direction: {100} imperfect, {110} indistinct, {111} and {011} parting

Habit: Crystals short prismatic, sometimes slender prismatic or pyramidal; granular, botryoidal or reniform masses

Fracture: Subconchoidal to uneven; brittle

Other: Alloyed to produce solder, pewter,, bronze, type-metal. Used as tin plate for cans and to produce casting metal

Comments: Occurs in granites and pegmatites associated with arsenopyrite, wolfram and copper minerals. Also occurs in alluvial deposits with ilmenite, monazite, zircon etc.

Optical Properties

Name: Cassiterite

Commodity: Tin (Sn)

Formula: SnO_2

Crystal System: Tetragonal

Color: Brownish Gray

Pleochroism: Distinct, brownish Gray to Gray

Anisotropism: Distinct, Gray, in oil

Internal Reflections: Very strong, yellow to yellowish brown

Reflectance 546nm: 11.5-12.44

Reflectance at 589nm: 11.3-12.2

Quant. Color Coordinates (QC):

Vickers Hardness Number (100g): 1168-1332, perfect

Polishing Hardness: Very high, greater than pyrite

Cleavage: May be visible, {100} imperfect, {110} indistinct, {111} and {011} parting

Form: Euhedral to subhedral prismatic crystals, often zoned and anhedral aggregates. Twinning common

Alteration:

Associated Minerals: With pyrite, stannite, arsenopyrite, wolframite, galena, sphalerite, rutile, hematite, magnetite, bismuth, bismuthinite, magnetite, pyrrhotite

Distinguishing Features: Internal reflections distinctive

Model: Occurs in granites and pegmatites associated with arsenopyrite, wolfram and copper minerals. Also occurs in alluvial deposits with ilmenite, monazite, zircon etc.

Physical Properties

Name: Celestite

Commodity: Strontium

Formula: SrSO4

Crystal System: Orthorhombic

Color: Colorless, white, grey, blue, green, yellow, red, brown

Opacity: Transparent to translucent

Luster: Vitreous

Streak:

SGLow: 3.97

SGHigh: 3.97

HardnessLow: 3

HardnessHigh: 3.5

Cleavage: 3

Direction: {001} perfect, {210} good, {010} indistinct

Habit: Crystals thin to thick tabular, or equant or pyramidal; nodules, fibrous veinlets; massive granular, lamellar, earthy

Fracture: Uneven; brittle

Other:

Comments:

Optical Properties

Name: Celestite

Group: Barite

Formula: SrSO4

Crystal System: Orthorhombic

Color: Colorless

Form: Euhedral to anhedral tabular crystals elongated

Relief: Fair, n>balsam

Birefringence: Weak, 0.009

2V: 51"

Nalpha or Nord.: 1.622

NBeta or Nextr.: 1.624

NGamma: 1.631

Optical Sign: Biaxial positive

Orientation: Length slow

Pleochroism:

Twinning:

Cleavage: Three perfect parallel to {001}, imperfect parallel to {110}

Extinction: Parallel to cleavage and outlines

Alteration:

Features: Sometimes fluoresces, name celestine also accepted. Similar to barite, axial angle is larger. Chhemical tests may be required.

Occurrence: Occurs in sedimentary limestones

Physical Properties

Name: Cerussite

Commodity: Lead (Pb)

Formula: $PbCO_3$

Crystal System: Orthorhombic

Color: Colorless, white, grey, smoky, blue, green black

Opacity: Transparent to subtranslucent

Luster: Adamantine, vitreous, resinous, pearly

Streak: Colorless to white

SGLow: 6.55

SGHigh: 6.55

HardnessLow: 3

HardnessHigh: 3.5

Cleavage: 4

Direction: {110} and {021} distinct, {010} and {012} trace

Habit: Crystals tabular, elongated along a-axis, or equant, acicular or thin tabular; twinned; granular, massive, jackstraw

Fracture: Conchoidal; brittle

Other: Pipes, alloys, batteries, radiation shielding, pigments

Comments: Associated with galena, anglesite, smithsonite and sphalerite in oxidized zone of lead deposits

Optical Properties

Name: Cerrusite

Group: Carbonate

Formula: $PbCO_3$

Crystal System: Orthorhombic

Color: Colorless

Form: Networks of fibers or columns, massive granular

Relief:

Birefringence: Strong, 0.273

2V: 8"-14"

Nalpha or Nord.: 1.803

NBeta or Nextr.: 2.074

NGamma: 2.076

Optical Sign: Biaxial negative

Orientation: Strong dispersion

Pleochroism:

Twinning:

Cleavage: Two, {110} distinct, {0,2,1} distinct

Extinction:

Alteration:

Features:

Occurrence: Associated with lead ores in limestone replacement deposits

Physical Properties

Name: Chalcocite

Commodity: Copper (Cu)

Formula: Cu2S

Crystal System: Monoclinic

Color: Blackish grey to black

Opacity: Opaque

Luster: Metallic

Streak: Blackish grey

SGLow: 5.5

SGHigh: 5.8

HardnessLow: 2.5

HardnessHigh: 3

Cleavage: 1

Direction: {110} indistinct

Habit: Crystals pseudohexagonal prisms formed by twinning; short prismatic or thick tabular; massive

Fracture: Conchoidal; brittle, somewhat sectile

Other: Alloyed with various metals to produce bronzes and brasses. Used in electrical industries where high electrical and thermal conductivity is required.

Comments: Chalcopyrite is the main ore of copper. In veins with other sulphides and with bornite and pyrite in porphyry copper deposits.

Optical Properties

Name: Chalcocite

Commodity: Copper (Cu)

Formula: Cu2S

Crystal System: Monoclinic

Color: Bluish white

Pleochroism: Very weak

Anisotropism: Weak to distinct, emerald green to light pinkish

Internal Reflections: None

Reflectance 546nm: 33.1-33.4

Reflectance at 589nm: 31.5-31.8

Quant. Color Coordinates (QC): 0.295, 0.304, 33.2 or 0.295, 0.304, 33.1 or 0.295, 0.303, 32.9

Vickers Hardness Number (100g): 84-87, perfect

Polishing Hardness: Greater than acanthite, equal to digenite, less than bornite

Cleavage: {110} indistinct

Form: Polyhedral aggregates and veins

Alteration: Exsolution intergrowths with bornite and other copper sulphides

Associated Minerals: With iron and copper sulphides

Distinguishing Features: Occurrence, polyhedral grains

Model: Chalcopyrite is the main ore of copper. In veins with other sulphides and with bornite and pyrite in porphyry copper deposits.

Physical Properties

Name: Chalcopyrite
Commodity: Copper (Cu)
Formula: CuFeS2
Crystal System: Tetragonal
Color: Brass-yellow, tarnishes iridescent
Opacity: Opaque
Luster: Metallic
Streak: Greenish black
SGLow: 4.35
SGHigh: 4.35
HardnessLow: 3.5
HardnessHigh: 4
Cleavage: 1
Direction: {011} sometimes distinct
Habit: Crystals sphenoidal, resemble tetrahedrons; massive, compact; reniform, botryoidal
Fracture: Uneven; brittle
Other: Alloyed with various metals to produce bronzes and brasses. Used in electrical industries where high electrical and thermal conductivity is required.
Comments: Chalcopyrite is the main ore of copper. In veins with other sulphides and with bornite and pyrite in porphyry copper deposits.

Optical Properties

Name: Chalcopyrite
Commodity: Copper (Cu)
Formula: CuFeS2
Crystal System: Tetragonal
Color: Yellow to brassy yellow, gold has greenish tinge, pyrite is more yellow
Pleochroism: Weak
Anisotropism: Weak, Gray blue to yellow green
Internal Reflections: None
Reflectance 546nm: 44.6-45.0
Reflectance at 589nm: 46.5-47.2
Quant. Color Coordinates (QC): 0.349, 0.369, 44.1
Vickers Hardness Number (100g): 187-203 or 181-192 basal section
Polishing Hardness: Equal to galena, less than sphalerite
Cleavage: {011} sometimes distinct
Form: Usually coarse to medium grained anhedral aggregates. Euhedral cystals rare. Twinning common. May contain laths of cubanite, stars of sphalerite, worms of pyrrhotite or mackinawite. Basketweav exsolution with bornite common
Alteration: Basketweav exsolution with bornite common. Alters to covellite along grain boundaries and cracks.
Associated Minerals: With pyrite, pyrrhotite, bornite, digenite, cubanite, sphalerite, galena, magnetite, pentlanddite, tetrahedrite etc.
Distinguishing Features: Color, weak anisotropy
Model: Chalcopyrite is the main ore of copper. In veins with other sulphides and with bornite and pyrite in porphyry copper deposits.

Physical Properties

Name: Chlorargyrite (Cerargyrite)

Commodity: Silver (Ag)

Formula: AgCl

Crystal System: Cubic

Color: grey, greenish grey, yellowish; may turn violet-brown on exposure to light

Opacity: Transparent to translucent

Luster: Resinous to adamantine, waxy

Streak:

SGLow: 5.556

SGHigh: 5.556

HardnessLow: 2.5

HardnessHigh: 2.5

Cleavage: 0

Direction:

Habit: Crystals cubic; massive, crusts and waxy coatings; columnar or fibrous

Fracture: Subconchoidal to uneven; sectile and ductile

Other: Used in coinage, jewellery, sterling ware, mirrors, electroplating, batteries and photographic and electronic products

Comments: Occurs in veins with native silver and cerrusite

Optical Properties

Name: Chlorargyrite (Cerargyrite)

Commodity: Silver (Ag)

Formula: AgCl

Crystal System: Cubic

Color: Gray, greenish Gray, yellowish; may turn violet-brown on exposure to light

Pleochroism: Transparent to translucent

Anisotropism: Resinous to adamantine, waxy

Internal Reflections:

Reflectance 546nm: 2.5

Reflectance at 589nm: 2.5

Quant. Color Coordinates (QC): 5.556

Vickers Hardness Number (100g): 5.556

Polishing Hardness: 0

Cleavage:

Form: Crystals cubic; massive, crusts and waxy coatings; columnar or fibrous

Alteration:

Associated Minerals:

Distinguishing Features: Crystals cubic; massive, crusts and waxy coatings; columnar or fibrous

Model: Occurs in veins with native silver and cerrusite

Physical Properties
Name: Chromite
Commodity: Chromium (Cr)
Formula: FeCr2O4
Crystal System: Cubic
Color: Black
Opacity: Opaque
Luster: Metallic
Streak: Brown
SGLow: 4.5
SGHigh: 4.8
HardnessLow: 5.5
HardnessHigh: 5.5
Cleavage: 0
Direction:
Habit: Crystals octahedral; massive, compact to fine granular
Fracture: Uneven; brittle
Other: Used as refractory in furnaces
Comments: Concentrated in ultramafic rocks where it forms layers in intrusions.

Optical Properties
Name: Chromite
Commodity: Chromium (Cr)
Formula: FeCr2O4
Crystal System: Cubic
Color: Dark Gray to brownish Gray
Pleochroism: None
Anisotropism: None, sometimes weak
Internal Reflections: Common, red-brown, none in Fe-rich varieties
Reflectance 546nm: 12.3
Reflectance at 589nm: 12.1
Quant. Color Coordinates (QC): 0.304, 0.309, 12.2
Vickers Hardness Number (100g): 1332, perfect
Polishing Hardness: Greater than skutterudite and arsenopyrite, less than pyrite
Cleavage:
Form: Subhedral to euhedral crystals and aggregates. Often zoned with lighter Fe-rich rims. Exsolution of hematite, ilmenitemagnetite, rutile and ulvospinel infrequent
Alteration: Exsolution of hematite, ilmenitemagnetite, rutile and ulvospinel infrequent
Associated Minerals: With magnetite, ilmenite, platinum, pentlandite, pyrrhotite and millerite
Distinguishing Features: Crystals octahedral; massive, compact to fine granular, internal reflections and occurrence distinctive
Model: Concentrated in ultramafic rocks where it forms layers in intrusions.

Physical Properties

Name: Chrysocolla
Commodity: Copper (Cu)
Formula: Cu2H2Si2O5(OH)4
Crystal System: Monoclinic
Color: Green, bluish green, blue
Opacity: Transluscent
Luster: Vitreous, greasy, dull
Streak: White to pale blue or green
SGLow: 2
SGHigh: 2.2
HardnessLow: 2
HardnessHigh: 4
Cleavage: None
Direction: None
Habit: Reniform, botryoidal masses
Fracture: Uneven to conchoidal
Other: Minor ore of copper, ornamental stone
Comments: Occurs in the oxide zone of copper deposits

Optical Properties

Name: Chrysocolla
Commodity: Copper
Formula: Cu2H2Si2O5(OH)4
Crystal System: Monoclinic
Color:
Pleochroism:
Anisotropism:
Internal Reflections:
Reflectance 546nm:
Reflectance at 589nm:
Quant. Color Coordinates (QC):
Vickers Hardness Number (100g):
Polishing Hardness:
Cleavage:
Form:
Alteration:
Associated Minerals:
Distinguishing Features:

Physical Properties

Name: Chrysotile (subgroup name of Serpintinites)

Commodity: Asbestos

Formula: Mg3Si2O5(OH)4

Crystal System: Monoclinic

Color: Yellow, white, grey, green

Opacity: Opaque

Luster: Non metallic

Streak:

SGLow: 2.55

SGHigh: 2.56

HardnessLow: 2.5

HardnessHigh: 2.5

Cleavage: 1

Direction: {001} perfect

Habit: Fibrous, tabular

Fracture: Plates

Other: Heat resistant, used for brake pads and fire retardent materials, roofing and cladding and fire retardent paints

Comments: Associated with serpentinite where it forms an alteration product

Optical Properties

Name: Chrysotile

Group: Serpentine

Formula: Mg3Si2O5(OH)4

Crystal System: Monoclinic

Color: Colorless

Form: Cross fibre veinlets (asbestiform)

Relief: Low slightly > balsam

Birefringence: Moderate, 0.011-0.014

2V: 0"-50"

Nalpha or Nord.: 1.493-1.546

NBeta or Nextr.: 1.504-1.550

NGamma: 1.517-1.557

Optical Sign: Biaxial,positive

Orientation: Length slow

Pleochroism:

Twinning:

Cleavage: One perfect {001}

Extinction: Parallel to length

Alteration:

Features: Asbestiform structure distinctive.

Occurrence: A metamorphic mineral principally found in serpentinites

Physical Properties

Name: Cinnabar

Commodity: Mercury (Hg)

Formula: HgS

Crystal System: Trigonal

Color: Scarlet, brownish red, brown, black, lead-grey

Opacity: Transparent to translucent

Luster: Adamantine, submetallic to dull

Streak: Scarlet to reddish brown

SGLow: 8.09

SGHigh: 8.09

HardnessLow: 2

HardnessHigh: 2.5

Cleavage: 1

Direction: {10-10} perfect

Habit: Crystals rhombohedral to thick tabular, stout to slender prismatic; crystalline crusts, powdery coatings

Fracture: Conchoidal to uneven; brittle; slightly sectile

Other: Used in scientific instruments, electrical goods, manufacture of paint, drugs and chemicals

Comments: Occurs in veins and fractures in metamorphic in sedimentary rocks in metamorphic belts. Only ore of mercury

Optical Properties

Name: Cinnabar

Commodity: Mercury (Hg)

Formula: HgS

Crystal System: Trigonal, hexagonal

Color: White with bluish Gray tint

Pleochroism: Distinct in oil

Anisotropism: Distinct in oil, masked by internal reflections

Internal Reflections: Very strong, red

Reflectance 546nm: 24.6-29.6

Reflectance at 589nm: 23.8-28.2

Quant. Color Coordinates (QC): 0.297, 0.302, 24.5

Vickers Hardness Number (100g): 935-1131

Polishing Hardness: Greater than antimony, less than galena, pyrite

Cleavage: {10-10} perfect

Form: Euhedral to subhedral crystals and aggregates

Alteration:

Associated Minerals: With pyrite, marcasite, stibnite, chalcopyrite, tetrahedrite, bornite, gold, realgar, orpiment, galena, enargite, cassiterite

Distinguishing Features: Crystals rhombohedral to thick tabular, stout to slender prismatic; crystalline crusts, powdery coatings, internal reflections distinctive

Model: Occurs in veins and fractures in metamorphic in sedimentary rocks in metamorphic belts. Only ore of mercury

Physical Properties

Name: Clausthalite

Commodity: Selenium (Se)

Formula: PbSe

Crystal System: Cubic

Color: Bright lead-grey

Opacity: Opaque

Luster: Metallic

Streak:

SGLow: 8.08

SGHigh: 8.22

HardnessLow: 2.5

HardnessHigh: 3

Cleavage: 1

Direction: {001} good

Habit: Massive, fine-grained, foliated

Fracture: Granular; brittle

Other: Used in fade resistant paint pigments, glass production, chemical industry and photo-electric instruments. Alloyed with copper and steel to improve workability

Comments: Selenium minerals are associated with sulphides in metal sulphide ore deposits. Recovered as a byproduct during copper sulphide ore smelting

Optical Properties

Name: Clausthalite

Commodity: Selenium (Se)

Formula: PbSe

Crystal System: Cubic

Color:

Pleochroism:

Anisotropism:

Internal Reflections:

Reflectance 546nm:

Reflectance at 589nm:

Quant. Color Coordinates (QC):

Vickers Hardness Number (100g):

Polishing Hardness:

Cleavage: {001} good

Form: Massive, fine-grained, foliated

Alteration:

Associated Minerals:

Distinguishing Features:

Model: Selenium minerals are associated with sulphides in metal sulphide ore deposits. Recovered as a byproduct during copper sulphide ore smelting

Physical Properties

Name: Cobaltite

Commodity: Cobalt (Co)

Formula: CoAsS

Crystal System: Orthorhombic ps. cub.

Color: Tin-white, steel-grey with violet tint, greyish black

Opacity: Opaque

Luster: Metallic, brilliant to dull

Streak: Greyish black

SGLow: 6.33

SGHigh: 6.33

HardnessLow: 5.5

HardnessHigh: 5.5

Cleavage: 1

Direction: {001} perfect

Habit: Crystals have isometric forms; massive, granular to compact

Fracture: Uneven; brittle

Other: Alloy to produce high temperature steels and magnetic alloys. Used as a catalyst in the chemical industries and as cement in sintered carbide cutting equipment

Comments: Occurs in veins with arsenopyrite, silver, calcite and nickel minerals. Cobalt is extracted as a byproduc during lead, copper and nickel smelting

Optical Properties

Name: Cobaltite

Commodity: Cobalt (Co)

Formula: CoAsS

Crystal System: Orthorhombic ps. cub.

Color: White with pink or violet tint

Pleochroism: Weak, white to pinkish

Anisotropism: Weak to distinct in oil, blue Gray to brown

Internal Reflections: None

Reflectance 546nm: 50.5

Reflectance at 589nm: 51.9

Quant. Color Coordinates (QC): 0.320, 0.326, 50.7

Vickers Hardness Number (100g): 935-1131

Polishing Hardness: Greater than skutterudite and arsenopyrite, less tha pyrite

Cleavage: May be visible {001} perfect

Form: Euhedral to subhedral isometric crystals and aggregates. Twinning and zoning may be present

Alteration: May be zoned

Associated Minerals: With niccolite, silver, gold, chalcopyrite, arsenopyrite, bismuth or uraninite, Ni-Co arsenides

Distinguishing Features: Crystals have isometric forms; massive, granular to compact, weak anistropy

Model: Occurs in veins with arsenopyrite, silver, calcite and nickel minerals. Cobalt is extracted as a byproduc during lead, copper and nickel smelting

Physical Properties

Name: Colemanite

Commodity: Borates

Formula: Ca2B6O11+5H2O

Crystal System: Monoclinic

Color: Colorless, white, yellowish white, greyish

Opacity: Transparent to translucent

Luster:

Streak:

SGLow: 2.42

SGHigh: 2.42

HardnessLow: 4.5

HardnessHigh: 4.5

Cleavage: 2

Direction: {010} perfect, {001} distinct

Habit: Crystals equant, short prismatic, pseudohexagonal; massive; rounded aggregates

Fracture: Subconchoidal to uneven; brittle

Other: Used in manufacture of insulating fibreglass, as fluxes for manufacture of glass and enamels. Borax is used in cloth manufacture and tanning, soap and glue, as preservatives and antiseptics

Comments: Lines cavities in sedimentary rocks and in evaporite basins with borax

Optical Properties

Name: Colemantite

Group: Borate

Formula: Ca2B6O11+5H2O

Crystal System: Monoclinic

Color: Colorless, milky white, yellowish, grey or muddy

Form: Short, prismatic crystals, massive and compact

Relief: Moderate to high, n>balsam

Birefringence: Moderate, 0.028

2V: 55"

Nalpha or Nord.: 1.586

NBeta or Nextr.: 1.592

NGamma: 1.614

Optical Sign: Biaxial positive

Orientation: Dispersion slight r>v, optic plane at right angles to {010}

Pleochroism:

Twinning:

Cleavage: Two {010} perfect, {001} fair

Extinction:

Alteration:

Features: Slightly soluble in water, indices of refraction > kernite and borax

Occurrence: Occurs within beds in desert environments

Physical Properties

Name: Coltan (Columbite-Tantalite)

Commodity: Niobium (Nb)

Formula: (Fe,Mn)Ta2O6

Crystal System: Orthorhombic

Color: Black to brownish black, tarnishes iridescent

Opacity: Opaque

Luster: Submetallic to weakly vitreous

Streak:

SGLow: 8.2

SGHigh: 8.2

HardnessLow: 6

HardnessHigh: 6.5

Cleavage: 2

Direction: {010} distinct, {100} less distinct

Habit: Crystals thin to thick tabular, short prismatic, equant or pyramidal; aggregates of parallel to divergent crystals

Fracture: Subconchoidal to uneven; brittle

Other: Important ferror-alloy to inhibit intergranular corrosion at high temperatures. Important metal used in electronics.

Comments: Associated with cassiterite, wolframite, spodumene and tourmaline in granite pegmatites

Optical Properties

Name: Coltan (Columbite-Tantalite)

Commodity: Niobium (Nb)

Formula: (Fe,Mn)Ta2O6

Crystal System: Orthorhombic

Color: Black to brownish black, tarnishes iridescent

Pleochroism: Opaque

Anisotropism: Submetallic to weakly vitreous

Internal Reflections:

Reflectance 546nm: 6

Reflectance at 589nm: 6.5

Quant. Color Coordinates (QC): 8.2

Vickers Hardness Number (100g): 8.2

Polishing Hardness: 2

Cleavage: {010} distinct, {100} less distinct

Form: Crystals thin to thick tabular, short prismatic, equant or pyramidal; aggregates of parallel to divergent crystals

Alteration:

Associated Minerals:

Distinguishing Features: Crystals thin to thick tabular, short prismatic, equant or pyramidal; aggregates of parallel to divergent crystals

Model: Associated with cassiterite, wolframite, spodumene and tourmaline in granite pegmatites

Physical Properties

Name: Copper

Commodity: Copper (Cu)

Formula: Cu

Crystal System: Cubic

Color: Copper-red, brown; pale rose when fresh, pale pink

Opacity: Opaque

Luster: Metallic

Streak: Pale red

SGLow: 8.94

SGHigh: 8.94

HardnessLow: 2.5

HardnessHigh: 3

Cleavage: 0

Direction:

Habit: Crystals cubic, octahedral, dodecahedral, tetrahexahedral; arborescent, wirelike, massive, powdery

Fracture: Hackly; malleable and ductile

Other: Alloyed with various metals to produce bronzes and brasses. Used in electrical industries where high electrical and thermal conductivity is required.

Comments: Occurs in small amounts with other copper minerals

Optical Properties

Name: Copper

Commodity: Copper (Cu)

Formula: Cu

Crystal System: Cubic

Color: Pink with brown tarnish

Pleochroism: None

Anisotropism: Isotropic, scratches appear anisotropic

Internal Reflections: None

Reflectance 546nm: 60.6

Reflectance at 589nm: 87.0

Quant. Color Coordinates (QC):

Vickers Hardness Number (100g): 96-104, perfect

Polishing Hardness: Greater than chalcocite, less than cuprite

Cleavage:

Form: Fine to coarse grained aggregates, dendritic or spearlike crystals infrequent. Zoning with silver or arsenic infrequent. Lamellar twinning visible when etched.

Alteration:

Associated Minerals: With cuprite, chalcocite, chalcopyrite, enargite, bornite, pyrrhotite, iron and magnetite

Distinguishing Features: Crystals cubic, octahedral, dodecahedral, tetrahexahedral; arborescent, wirelike, massive, powdery, color and anisotropy

Model: Occurs in small amounts with other copper minerals

Physical Properties

Name: Corundum

Group: Corundum

Formula: Al2O3

Crystal System: Trigonal

Color: Colorless, grey, brown, blue, red, green, yellow, orange, purple

Opacity: Transparent to translucent

Luster: Vitreous to adamantine

Streak: Colorless

SGLow: 4.0

SGHigh: 4.1

HardnessLow: 9

HardnessHigh: 9

Cleavage: 2

Direction: {0001} and {10-11} parting

Habit: Crystals often well-developed, steep-pyramidal, prismatic, tabular; barrel-shaped; massive

Fracture: Conchoidal

Other: Cryptocrystalline variety of quartz

Comments: Sometimes fluoresces under SW and LW

Optical Properties

Name: Corundum

Group: Corundum

Formula: Al2O3

Crystal System: Trigonal

Color: Colorless

Form: Euhedral tabular to prismatic cystals common, cross sections are six sided and may show zoning.

Relief: Very high, n>balsam

Birefringence: Weak, 0.008-0.009

2V: 4.1

Nalpha or Nord.: 1.759 to 1.763

NBeta or Nextr.: 1.767 to 1.772

NGamma:

Optical Sign: Uniaxial neg

Orientation: Length slow (tabular), length fast (prismatic)

Pleochroism: May be pleochroic in thick sections

Twinning: Twinning lamellae wit {101_1} as twin plane

Cleavage: Two {0001} and {10-11} parting. Parting oftem parallel to rhombohedron {10_11} or pinacoid {0001} or both

Extinction:

Alteration: Unusual

Features: Very high relief with weak birefringence, parting and twin lamellae

Occurrence: Cryptocrystalline variety of quartz, Characteristic of corundum syenites, contact metamorphic limestones and metamorphic shales

Physical Properties
Name: Covellite
Commodity: Copper (Cu)
Formula: CuS
Crystal System: Hexagonal
Color: Light to dark indigo-blue
Opacity: Opaque
Luster: Submetallic to dull
Streak: Shining grey-black
SGLow: 4.681
SGHigh: 4.681
HardnessLow: 1.5
HardnessHigh: 2
Cleavage: 1
Direction: {0001} perfect
Habit: Crystals thin tabular hexagonal plates; massive, foliated
Fracture: Uneven; brittle, thin laminae flexible
Other: Alloyed with various metals to produce bronzes and brasses. Used in electrical industries where high electrical and thermal conductivity is required.
Comments: Chalcopyrite is the main ore of copper. In veins with other sulphides and with bornite and pyrite in porphyry copper deposits.

Optical Properties
Name: Covellite
Commodity: Copper (Cu)
Formula: CuS
Crystal System: Hexagonal
Color: Indigo blue with violet tint to bluish white
Pleochroism: Purple to violet red, to blue Gray in oil
Anisotropism: Extreme, re orange to brownish
Internal Reflections: None
Reflectance 546nm: 7.2-23.7
Reflectance at 589nm: 4.2-21.2
Quant. Color Coordinates (QC): 0.224, 0.226, 6.8
Vickers Hardness Number (100g): 128-138, slightly fractured
Polishing Hardness: Less than chalcopyrite
Cleavage: {0001} perfect
Form: Lathlike and platelike crystals, subhedral to anhedral aggregates
Alteration: Appears as alteration laths on copper and other iron sulphides
Associated Minerals: With pyrite, chalcopyrite, bornite and other copper sulphide minerals
Distinguishing Features: Crystals thin tabular hexagonal plates; massive, foliated, Blue color, strong pleochroism and anisotropy distinctive
Model: Chalcopyrite is the main ore of copper. In veins with other sulphides and with bornite and pyrite in porphyry copper deposits.

Physical Properties

Name: Cuprite

Commodity: Copper (Cu)

Formula: Cu2O

Crystal System: Cubic

Color: Red, brownish red, purplish red to almost black

Opacity: Transparent to translucent

Luster: Adamantine or submetallic to earthy

Streak: Brownish red, shining

SGLow: 6.14

SGHigh: 6.14

HardnessLow: 3.5

HardnessHigh: 4

Cleavage: 2

Direction: {111} interrupted, {001} rare

Habit: Crystals cubic, octahedral or dodecahedral; hair-like forming wads or mats (chalcotrichite); massive, compact, granular

Fracture: Conchoidal to uneven; brittle

Other: Alloyed with various metals to produce bronzes and brasses. Used in electrical industries where high electrical and thermal conductivity is required.

Comments: Chalcopyrite is the main ore of copper. Cuprite is found in the oxidized zone of copper deposits

Optical Properties

Name: Cuprite

Commodity: Copper (Cu)

Formula: Cu2O

Crystal System: Cubic

Color: Light bluish Gray in air, more blue in oil

Pleochroism: Very weak

Anisotropism: Strong anomolous colors from Gray blue to olive green

Internal Reflections: Deep red

Reflectance 546nm: 26.6

Reflectance at 589nm: 24.6

Quant. Color Coordinates (QC): 0.287, 0.300, 26.3

Vickers Hardness Number (100g): 193-207

Polishing Hardness: Greater than chalcopyrite, less than pyrrhotite

Cleavage: {111} interrupted, {001} rare

Form: Crystals cubic, octahedral or dodecahedral; hair-like forming wads or mats (chalcotrichite); massive, compact, granular

Alteration: Replaces copper sulphides

Associated Minerals: With goethite, tenorite, pyrite, marcasite, delafossite

Distinguishing Features: Crystals cubic, octahedral or dodecahedral; hair-like forming wads or mats (chalcotrichite); massive, compact, granular. Anomolous anisotropy and deep red internal reflections are distinctive

Model: Chalcopyrite is the main ore of copper. Cuprite is found in the oxidized zone of copper deposits

Physical Properties

Name: Diamond

Commodity: Diamond

Formula: C

Crystal System: Cubic

Color: Colorless, white to blue-white, grey, yellow, brown, green, red, black

Opacity: Transparent to opaque

Luster: Adamantine to greasy

Streak:

SGLow: 3.51

SGHigh: 3.51

HardnessLow: 10

HardnessHigh: 10

Cleavage: 1

Direction: {111} perfect

Habit: Crystals octahedral, or cubic, dodecahedral, tetrahedral; flattened, faces curved or striated, spherical aggregates

Fracture: Conchoidal; brittle

Other: Gemstone and used as as an abrasive and in cutting tools

Comments: Occurs in lamproites and kimberlites and in alluvial deposits derived from these rocks

Optical Properties

Name: Diamond

Group: Diamond

Formula: C

Crystal System: Isometric

Color: Colorless

Form: Euhedral crystals, also granular

Relief: High

Birefringence: None

2V: None

Nalpha or Nord.: 2.4175-2.4178

NBeta or Nextr.:

NGamma:

Optical Sign:

Orientation:

Pleochroism:

Twinning:

Cleavage: None

Extinction: Isotropic

Alteration:

Features:

Occurrence: Associated with lamproite and kimberlite intrusions

Physical Properties

Name: Diaspore

Commodity: Aluminum (Al)

Formula: AlO(OH)

Crystal System: Orthorhombic

Color: White to colorless, yellowish, greenish, lilac, pink, brownish

Opacity: Transparent to subtranslucent

Luster: Vitreous, pearly on cleavages

Streak:

SGLow: 3.3

SGHigh: 3.5

HardnessLow: 6.5

HardnessHigh: 7

Cleavage: 3

Direction: {010} perfect, {110} distinct, {100} traces

Habit: Crystals thin elongated plates; acicular or tabular, striated; massive, foliated, scaly, stalactitic

Fracture: Conchoidal; brittle

Other: Manufacture of Alumininum Alloys

Comments: Mixed with gibbsite, boehmite, iron oxides and silica to form bauxite deposits. Bauxite is pisolitic to massive and forms on the surface from weathering of granitic and gneissic rocks in tropical and subtropical climates

Optical Properties

Name: Diaspore

Group: Bauxite

Formula: AlO(OH)

Crystal System: Orthorhombic

Color: Colorless to pale blue

Form: Tabular parallel to {010}. Also occurs in mineral aggregates

Relief: High, n>balsam

Birefringence: Strong 0.048

2V: 84"

Nalpha or Nord.: 1.702

NBeta or Nextr.: 1.722

NGamma: 1.750

Optical Sign: Biaxial positive

Orientation: Length fast, dispersion, r<v weak

Pleochroism: May be pleochroic in thick sections

Twinning:

Cleavage: Perfect, one direction {010}

Extinction: Parallel

Alteration: Dimorphous with boehmite

Features: Cleavage, crystals and relief

Occurrence: Occurs in metamorphic schists, altered igneous rocks and flint clays

Physical Properties

Name: Dolomite

Commodity: Magnesium (Mg)

Formula: $CaMg(CO_3)_2$

Crystal System: Trigonal

Color: Colorless, white, greyish, greenish, pale brown, pinkish

Opacity: Transparent to subtranslucent

Luster: Vitreous to pearly

Streak:

SGLow: 2.85

SGHigh: 2.85

HardnessLow: 3.5

HardnessHigh: 4

Cleavage: 1

Direction: {10-11} perfect

Habit: Crystals simple rhombohedrons, often with curved faces, rarely tabular or octahedral; crystalline aggregates; massive

Fracture: Subconchoidal; brittle

Other: Alloyed with aluminum to im,prove strength and corrosion resistance. Manufacture of Iron nodules and desulphuring. Used in petrol, zirconium and titanium production and magnesium die-castings

Comments: Occurs in sedimentary basins as rock forming mineral and in veins with galena and sphalerite

Optical Properties

Name: Dolomite

Group: Dolomite

Formula: $CaMg(CO_3)_2$

Crystal System: Hexagonal

Color: Colorless to grey

Form: Fine to coarse subhedral grains. Euhedral, rhombohedral grains. Zonal structure common duet variation in iron content.

Relief: High in long direction of rhomb, low in short direction.

Birefringence: Extreme, 0.180-0.190

2V:

Nalpha or Nord.: 1.500-1.526

NBeta or Nextr.: 1.680-1.716

NGamma:

Optical Sign: Uniaxial negative

Orientation:

Pleochroism:

Twinning: Polysynthetic twinning along {022_1}

Cleavage: One perfect rhombohedral parallel to {101_1}

Extinction: Symmetrical to cleavage traces, curved crsytals show wavy extinction

Alteration:

Features: Similar to calcite and magnesite. Distinguished from calcite by euhedral crystal form, zonal structure and twinning lamellae parallel to short diagonal. Staining may be required to differentiate magnesite.

Occurrence: The major constituent of sedimentary and metamorphic dolomite rock. Also occurs in veins and hydrothermal deposits

Physical Properties

Name: Enargite

Commodity: Copper (Cu)

Formula: Cu3AsS4

Crystal System: Orthorhombic

Color: Greyish black to iron-black

Opacity: Opaque

Luster: Metallic

Streak: Greyish black

SGLow: 4.45

SGHigh: 4.45

HardnessLow: 3

HardnessHigh: 3

Cleavage: 4

Direction: {110} perfect, {100} and {010} distinct, {001} indistinct

Habit: Crystals prismatic or tabular; massive, granular or prismatic

Fracture: Uneven; brittle

Other: Alloyed with various metals to produce bronzes and brasses. Used in electrical industries where high electrical and thermal conductivity is required.

Comments: Chalcopyrite is the main ore of copper. In veins with other sulphides and with bornite and pyrite in porphyry copper deposits.

Optical Properties

Name: Enargite

Commodity: Copper (Cu)

Formula: Cu3AsS4

Crystal System: Orthorhombic

Color: Pinkish Gray to pinkish brown, darker in oil

Pleochroism: Distinct in oil, Grayish pink to pinkish Gray to Grayish violet

Anisotropism: Strong, blue, green, red, orange

Internal Reflections: May show deep red

Reflectance 546nm: 24.2-25.2

Reflectance at 589nm: 23.8-25.7

Quant. Color Coordinates (QC): 4.45

Vickers Hardness Number (100g): 285-327

Polishing Hardness: Greater than galena, chalcocite, bornite, less than sphalerite, equal to tennantite

Cleavage: Often seen, {110} perfect, {100} and {010} distinct, {001} indistinct

Form: Anhedral to subhedral grains. Crystals prismatic or tabular; massive, granular or prismatic

Alteration:

Associated Minerals: With pyrite, chalcopyrite, bornite, sphalerite, tennantite, galena, chalcocite, covellite, arsenopyrite

Distinguishing Features: Crystals prismatic or tabular; massive, granular or prismatic. Pleochroism and anisotropy distinctive

Model: Chalcopyrite is the main ore of copper. In veins with other sulphides and with bornite and pyrite in porphyry copper deposits.

Physical Properties

Name: Eucairite

Commodity: Selenium (Se)

Formula: CuAgSe

Crystal System: Orthorhombic

Color: Brilliant silver-white and lead-grey, tarnishes bright bronze

Opacity: Opaque

Luster: Metallic

Streak: Shining

SGLow: 7.6

SGHigh: 7.8

HardnessLow: 2.5

HardnessHigh: 2.5

Cleavage: 0

Direction:

Habit: Massive, granular; thin films on calcite

Fracture: Subconchoidal to uneven; brittle

Other: Used in fade resistant paint pigments, glass production, chemical industry and photo-electric instruments. Alloyed with copper and steel to improve workability

Comments: Selenium minerals are associated with sulphides in metal sulphide ore deposits. Recovered as a byproduct during copper sulphide ore smelting

Optical Properties

Name: Eucairite

Commodity: Selenium (Se)

Formula: CuAgSe

Crystal System: Orthorhombic

Color:

Pleochroism:

Anisotropism:

Internal Reflections:

Reflectance 546nm:

Reflectance at 589nm:

Quant. Color Coordinates (QC):

Vickers Hardness Number (100g):

Polishing Hardness:

Cleavage:

Form: Massive, granular; thin films on calcite

Alteration:

Associated Minerals:

Distinguishing Features: Massive, granular; thin films on calcite

Model: Selenium minerals are associated with sulphides in metal sulphide ore deposits. Recovered as a byproduct during copper sulphide ore smelting

Physical Properties

Name: Ferrotantalite

Commodity: Tantalum (Ta)

Formula: FeTa2O6

Crystal System: Orthorhombic

Color: Black to brownish black, tarnishes iridescent

Opacity: Opaque

Luster: Submetallic to weakly vitreous

Streak:

SGLow: 8.2

SGHigh: 8.2

HardnessLow: 6

HardnessHigh: 6.5

Cleavage: 2

Direction: {010} distinct, {100} less distinct

Habit: Crystals thin to thick tabular, short prismatic, equant or pyramidal; aggregates of parallel to divergent crystals

Fracture: Subconchoidal to uneven; brittle

Other: Very high corrosion resistance. Used in production of special steels for medical use, in chemical and electrical processes, electrodes, cutting tools and manufacture of electrodes

Comments: Associated with cassiterite, wolframite, spodumene and tourmaline in granite pegmatites. Tantalum is also recovered from tin slags as a byproduct

Optical Properties

Name: Ferrotantalite

Commodity: Tantalum (Ta)

Formula: FeTa2O6

Crystal System: Orthorhombic

Color:

Pleochroism:

Anisotropism:

Internal Reflections:

Reflectance 546nm:

Reflectance at 589nm:

Quant. Color Coordinates (QC):

Vickers Hardness Number (100g):

Polishing Hardness:

Cleavage: {010} distinct, {100} less distinct

Form: Crystals thin to thick tabular, short prismatic, equant or pyramidal; aggregates of parallel to divergent crystals

Alteration:

Associated Minerals:

Distinguishing Features:

Model: Associated with cassiterite, wolframite, spodumene and tourmaline in granite pegmatites. Tantalum is also recovered from tin slags as a byproduct

Physical Properties

Name: Fluorapatite

Commodity: Phosphates

Formula: Ca5(PO4)3F

Crystal System: Hexagonal

Color: Colorless, white, grey, yellow to yellowish green, green, blue, violet, red, brown

Opacity: Transparent to opaque

Luster: Vitreous to subresinous, silky

Streak: White

SGLow: 3.1

SGHigh: 3.2

HardnessLow: 5

HardnessHigh: 5

Cleavage: 2

Direction: {0001} indistinct, {10-10} trace

Habit: Crystals short to long prismatic, thin to thick tabular or complex; massive compact to coarse granular; fibrous

Fracture: Conchoidal to uneven; brittle

Other: Fertiliser

Comments: Formed by precipitation from sea water, replacement of limestones or shelly marine sediments, in guano deposits and accumulation of grains from pre-existing rocks. Major component of phosphorite rocks

Optical Properties

Name: Fluorapatite (Apatite)

Group: Sillimanite

Formula: Ca5(PO4)3F

Crystal System: Hexagonal

Color: Colorless

Form: Crystals short to long prismatic, thin to thick tabular or complex; massive compact to coarse granular; fibrous

Relief: Moderate, n>balsam

Birefringence: Weak, 0.003-0.004

2V:

Nalpha or Nord.: 1.630-1.651

NBeta or Nextr.: 1.633-1.655

NGamma:

Optical Sign: Uniaxial negative

Orientation: Length fast, tabular sections are length slow

Pleochroism:

Twinning:

Cleavage: Two, imperfect basal {0001} may show as cross fractures, {101_0} may show parallel to length

Extinction: Parallel

Alteration:

Features: Often fluoresces, phosphorescent in UV light, or thermoluminescent. Hexagonal form distinctive

Occurrence:

Physical Properties

Name: Fluorite

Commodity: Flourspar

Formula: CaF_2

Crystal System: Cubic

Color: Colorless, purple, blue, green, yellow, white, pink, red, brown, bluish black

Opacity: Transparent to translucent

Luster: Vitreous

Streak:

SGLow: 3.18

SGHigh: 3.18

HardnessLow: 4

HardnessHigh: 4

Cleavage: 1

Direction: {111} perfect

Habit: Crystals cubes or octahedrons, rarely dodecahedrons; massive, coarse to fine granular; botryoidal, fibrous

Fracture: Subconchoidal to splintery; brittle

Other: Used as a flux in steel making, manufacture of optical equipment, production of hydroflouric acid and flourocarbons and as gemstone

Comments: In hydrothermal veins and replacement deposits

Optical Properties

Name: Fluorite

Group: Fluorite

Formula: CaF_2

Crystal System: Isometric, cubic

Color: Colorless, purple bands or sppots

Form: Sometimes euhedral with cubic outlines, usually anhedral

Relief: Fairly high, n<balsam

Birefringence: Nil, dark between crossed polars

2V:

Nalpha or Nord.: 1.434

NBeta or Nextr.:

NGamma:

Optical Sign:

Orientation:

Pleochroism:

Twinning:

Cleavage: One {111} perfect. Perfect octohedral cleavage. Usually visble as two intersecting lines at angles of 70" and 110", or three intersecting ;lines at 60" and 120"

Extinction:

Alteration:

Features: High relief, perfect octohedral cleavage, isometric, purple spots or bands.

Occurrence: Mostly in veins associated with granites

Physical Properties

Name: Franklinite
Commodity: Zinc (Zn)
Formula: (Zn,Mn,Fe)(Fe,Mn)2O4
Crystal System: Cubic
Color: Iron-black
Opacity: Opaque
Luster: Metallic to dull
Streak: Reddish brown to black
SGLow: 5.07
SGHigh: 5.22
HardnessLow: 5.5
HardnessHigh: 6.5
Cleavage: 1
Direction: {111} parting, fair
Habit: Crystals octahedral, large, may be modified or rounded; massive, compact, coarse to fine granular
Fracture: Subconchoidal to uneven; brittle
Other: Alloyed to make brass and zinc die castings. Corrosion resistant coatings on steel (galvanizing) and iron. Manufacture of corrosion resitant paints, pigments and fillers etc.
Comments: Associated with zincite and willemite in contact metamorposed limestones

Optical Properties

Name: Franklinite
Commodity: Zinc (Zn)
Formula: (Zn,Mn,Fe)(Fe,Mn)2O4
Crystal System: Cubic
Color: White
Pleochroism:
Anisotropism:
Internal Reflections:
Reflectance 546nm:
Reflectance at 589nm:
Quant. Color Coordinates (QC):
Vickers Hardness Number (100g):
Polishing Hardness:
Cleavage: {111} parting, fair
Form: Crystals octahedral, large, may be modified or rounded; massive, compact, coarse to fine granular
Alteration:
Associated Minerals:
Distinguishing Features:
Model: Associated with zincite and willemite in contact metamorposed limestones

Physical Properties

Name: Galena

Commodity: Lead (Pb)

Formula: PbS

Crystal System: Cubic

Color: Lead-grey

Opacity: Opaque

Luster: Bright metallic

Streak: Lead-grey

SGLow: 7.58

SGHigh: 7.58

HardnessLow: 2.5

HardnessHigh: 2.5

Cleavage: 1

Direction: {001} perfect

Habit: Crystals cubic, octahedral, or cuboctahedral, or tabular, skeletal, or reticulated; massive, cleavable; granular; fibrous

Fracture: Subconchoidal

Other: Pipes, alloys, batteries, radiation shielding, pigments

Comments: Main ore of lead. Occurs in veins with other sulphides, in pegmatites and as replacement deposits in limestones and dolomites. Often contains silver and is an important ore of this metal

Optical Properties

Name: Galena

Commodity: Lead (Pb)

Formula: PbS

Crystal System: Cubic

Color: White, occasionally pink tint

Pleochroism: None

Anisotropism: Isotropic, weak anomolous anisotropy may occur

Internal Reflections: None

Reflectance 546nm: 43.1

Reflectance at 589nm: 41.9

Quant. Color Coordinates (QC): 0.300, 0.304, 43.0

Vickers Hardness Number (100g): 59-65, perfect

Polishing Hardness: Greater than proustite, equal to chalcopyrite, less than tetrahedrite

Cleavage: {001} perfect, often seen with triangular pits

Form: Crystals cubic, octahedral, or cuboctahedral, or tabular, skeletal, or reticulated; massive, cleavable; granular; fibrous. Inclusions of tetrahedrite, Pb, Bi, silver, Pb-Sb sulfosalts and chalcopyrite. Occurs as inclusions in chalcopyrite, sphalerite

Alteration: Occurs as inclusions in chalcopyrite, sphalerite

Associated Minerals: Widely distributed with many different minerals

Distinguishing Features: Crystals cubic, octahedral, or cuboctahedral, or tabular, skeletal, or reticulated; massive, cleavable; granular; fibrous. Cubic crystals, triangular cleavage pits characteristic

Model: Main ore of lead. Occurs in veins with other sulphides, in pegmatites and as replacement deposits in limestones and dolomites. Often contains silver and is an important ore of this metal

Physical Properties
Name: Gallium
Commodity: Gallium (Ga)
Formula:
Crystal System:
Color:
Opacity:
Luster:
Streak:
SGLow:
SGHigh:
HardnessLow:
HardnessHigh:
Cleavage:
Direction:
Habit:
Fracture:
Other: To produce light emitting diodes and computer memory
Comments: A byproduct of aluminum smelting

Optical Properties
Name: Gallium
Commodity: Gallium (Ga)
Formula:
Crystal System:
Color:
Pleochroism:
Anisotropism:
Internal Reflections:
Reflectance 546nm:
Reflectance at 589nm:
Quant. Color Coordinates (QC):
Vickers Hardness Number (100g):
Polishing Hardness:
Cleavage:
Form:
Alteration:
Associated Minerals:
Distinguishing Features:
Model: A byproduct of aluminum smelting

Physical Properties

Name: Gibbsite

Commodity: Aluminum (Al)

Formula: Al(OH)3

Crystal System: Monoclinic

Color: White, greyish, greenish, reddish white

Opacity: Transparent to translucent

Luster: Vitreous, pearly on cleavage

Streak:

SGLow: 2.40

SGHigh: 2.40

HardnessLow: 2.5

HardnessHigh: 3.5

Cleavage: 1

Direction: {001} perfect

Habit: Crystals tabular, hexagonal to large size; massive, chalcedony-like coatings and crusts; stalactitic, concretionary, fibrous

Fracture: Tough

Other: Manufacture of Alumininum Alloys

Comments: Mixed with gibbsite, boehmite, iron oxides and silica to form bauxite deposits. Bauxite is pisolitic to massive and forms on the surface from weathering of granitic and gneissic rocks in tropical and subtropical climates

Optical Properties

Name: Gibbsite

Commodity: Aluminum (Al)

Formula: Al(OH)3

Crystal System: Monoclinic

Color: White, Grayish, greenish, reddish white

Pleochroism: Transparent to translucent

Anisotropism: Vitreous, pearly on cleavage

Internal Reflections:

Reflectance 546nm: 2.5

Reflectance at 589nm: 3.5

Quant. Color Coordinates (QC): 2.40

Vickers Hardness Number (100g): 2.40

Polishing Hardness: 1

Cleavage: {001} perfect

Form: Crystals tabular, hexagonal to large size; massive, chalcedony-like coatings and crusts; stalactitic, concretionary, fibrous

Alteration:

Associated Minerals:

Distinguishing Features: Crystals tabular, hexagonal to large size; massive, chalcedony-like coatings and crusts; stalactitic, concretionary, fibrous

Model: Mixed with gibbsite, boehmite, iron oxides and silica to form bauxite deposits. Bauxite is pisolitic to massive and forms on the surface from weathering of granitic and gneissic rocks in tropical and subtropical climates

Physical Properties

Name: Goethite

Commodity: Iron (Fe)

Formula: FeO(OH)

Crystal System: Orthorhombic

Color: Blackish brown, reddish or yellowish brown, brownish yellow

Opacity: Opaque

Luster: Adamantine-metallic to dull

Streak: Orange to brownish yellow

SGLow: 3.3

SGHigh: 4.3

HardnessLow: 5

HardnessHigh: 5.5

Cleavage: 2

Direction: {010} perfect, {100} distinct

Habit: Crystals prismatic, vertically striated, capillary or acicular in radiating clusters; massive radial, ocherous, colloform, compact

Fracture: Uneven; brittle

Other: Iron and Steel

Comments: Associated with hematite and limonite in plutonic rocks and Archean Banded Iron Formations and in duricrust over mafic and ultramafic bedrocks

Optical Properties

Name: Goethite

Commodity: Iron (Fe)

Formula: FeO(OH)

Crystal System: Orthorhombic

Color: Gray, with bluish tint

Pleochroism: Weak in air, distinct in oil, masked by internal reflections

Anisotropism: Distinct, Gray blue to Gray yellow to brownish

Internal Reflections: Brownish yellow to reddish brown

Reflectance 546nm: 15.5-17.5

Reflectance at 589nm: 15.0-16.6

Quant. Color Coordinates (QC): 0.291, 0.296, 17.5

Vickers Hardness Number (100g): 667

Polishing Hardness: Equal to lepidocrocite, less than magnetite, hematite

Cleavage: {010} perfect, {100} distinct

Form: Often in colloform bands, radiating fibrous crystals or pseudomorphs after pyrite. As veins, fracture fillings and botryoidal coatings

Alteration: Pseudomorphs after pyrite

Associated Minerals: With hematite, pyrite, lepidocrocite, pyrrhotite, manganese oxides, sphalerite, galena, chalcopyrite

Distinguishing Features: Crystals prismatic, vertically striated, capillary or acicular in radiating clusters; massive radial, ocherous, colloform, compact. Brownish yellow internal reflections distinctive

Model: Associated with hematite and limonitel in plutonic rocks and Archean Banded Iron Formations and in duricrust over mafic and ultramafic bedrocks. A secondary mineral.

Physical Properties

Name: Gold

Commodity: Gold (Au)

Formula: Au

Crystal System: Cubic

Color: Gold yellow; silver-white to orange-red from impurities

Opacity: Opaque

Luster: Metallic

Streak: Same as color

SGLow: 19.297

SGHigh: 19.297

HardnessLow: 2.5

HardnessHigh: 3

Cleavage: 0

Direction:

Habit: Crystals octahedral, dodecahedral or cubic, flattened and elongated, dendritic; massive as nuggets, grains or scales

Fracture: Hackly; malleable and ductile

Other: Monetary use, jewellery, electronics and for decoration

Comments: Occurs in quartz veins with sulphides and in alluvial deposits. Occurs in epithermal to hypothermal veins

Optical Properties

Name: Gold

Commodity: Gold (Au)

Formula: Au

Crystal System: Cubic

Color: Bright gold yellow; silver-white to orange-red from impurities

Pleochroism: None

Anisotropism: Isotropic

Internal Reflections: None

Reflectance 546nm: 71.5

Reflectance at 589nm: 83.4

Quant. Color Coordinates (QC): 0.384, 0.391, 72.7

Vickers Hardness Number (100g): 53-58, perfect

Polishing Hardness: Greater than galena, less than tetrahedrite, chalcopyrite

Cleavage:

Form: Usually as isolated grains and veinlets with sulphides

Alteration: Impurities of silver change color whitish

Associated Minerals: With many sulphides

Distinguishing Features: Gold yellow color, usually as isolated grains and veinlets with sulphides.

Model: Occurs in quartz veins with sulphides and in alluvial deposits. Occurs in epithermal to hypothermal veins

Physical Properties
Name: Goldamalgam
Commodity: Gold (Au)
Formula: (Au,Ag)Hg
Crystal System: Cubic
Color: Brass-yellow
Opacity:
Luster: Metallic
Streak:
SGLow: 15.47
SGHigh: 15.47
HardnessLow: 3
HardnessHigh: 3
Cleavage: 0
Direction:
Habit: Fine granular aggregates
Fracture: Conchoidal; malleable
Other: Monetary use, jewellery, electronics and for decoration
Comments: Occurs in quartz veins with sulphides and in alluvial deposits. Occurs in epithermal to hypothermal veins

Optical Properties
Name: Goldamalgam
Commodity: Gold (Au)
Formula: (Au,Ag)Hg
Crystal System: Cubic
Color: Brass-yellow
Pleochroism:
Anisotropism: None
Internal Reflections: None
Reflectance 546nm:
Reflectance at 589nm:
Quant. Color Coordinates (QC):
Vickers Hardness Number (100g):
Polishing Hardness:
Cleavage:
Form: Fine granular aggregates
Alteration:
Associated Minerals:
Distinguishing Features: Fine granular aggregates
Model: Occurs in quartz veins with sulphides and in alluvial deposits. Occurs in epithermal to hypothermal veins

Physical Properties

Name: Graphite

Commodity: Graphite

Formula: C

Crystal System: Hexagonal and trigonal

Color: Iron-black to steel-grey

Opacity: Opaque

Luster: Metallic, dull, earthy

Streak: Black

SGLow: 2.09

SGHigh: 2.23

HardnessLow: 1

HardnessHigh: 2

Cleavage: 1

Direction: {0001} perfect

Habit: Crystals thin tabular, hexagonal; massive, fine to coarse foliated; radiate aggregates, columnar, granular, scaly, earthy

Fracture: Flexible, inelastic; greasy feel; sectile

Other: Manufacture of crucibles, paint and foundry molds. Used as lubricant and electric furnace electrodes

Comments: Occurs in metamorphic rocks which have a source of carbon and as veins in igneous rocks and pegmatites

Optical Properties

Name: Graphite

Commodity: Graphite

Formula: C

Crystal System: Hexagonal and trigonal

Color: Grayish black

Pleochroism: Very strong, Grayish black to brownish Gray

Anisotropism: Very strong, straw yellow to brown or violet Gray

Internal Reflections: None

Reflectance 546nm: 6.8-17.4

Reflectance at 589nm: 7.0-18.0

Quant. Color Coordinates (QC):

Vickers Hardness Number (100g): 12-16, fractured

Polishing Hardness: Less than most minerals

Cleavage: Basal cleavage often visible{0001} perfect, straight extiction

Form: Plates, lathes and bladed aggregates. Straight extiction

Alteration:

Associated Minerals: Inclusions in pyrite, sphalerite, pyrrhotite and magnetite

Distinguishing Features: Crystals thin tabular, hexagonal; massive, fine to coarse foliated; radiate aggregates, columnar, granular, scaly, earthy

Model: Occurs in metamorphic rocks which have a source of carbon and as veins in igneous rocks and pegmatites. Occurs as isolated laths in many igneous and metamorphic rocks.

Physical Properties

Name: Grossular

Commodity: Garnet

Formula: Ca3Al2(SiO4)3

Crystal System: Cubic

Color: Colorless, white, grey, yellow to green, brown, pink, red, black

Opacity: Transparent to nearly opaque

Luster: Vitreous to resinous

Streak: White

SGLow: 3.4

SGHigh: 3.6

HardnessLow: 6.5

HardnessHigh: 7

Cleavage: 1

Direction: {110} parting

Habit: Crystals dodecahedrons or trapezohedrons; massive, compact; granular; embedded grains

Fracture: Conchoidal to uneven; brittle

Other: Abrasives, gemstones

Comments: Occurs in many types of metamorphic rocks and some igneous rocks and in alluvial and beach deposits

Optical Properties

Name: Grossular (Grossularite)

Group: Garnet

Formula: Ca3Al2(SiO4)3

Crystal System: Isometric, cubic

Color: Colorless to pale red, pale brown to brown, greenish

Form: Euhedral dodecahedrons in six sided trapezohedrons in eigth sided cross sections. Plygonal grains and aggregates.

Relief: V.high, n>balsam

Birefringence: Isotropic, but may show weak birefringence

2V:

Nalpha or Nord.: 1.739-1.763

NBeta or Nextr.:

NGamma:

Optical Sign:

Orientation:

Pleochroism:

Twinning:

Cleavage: Parting parallel to{110}. Irregular fractures

Extinction:

Alteration:

Features: Similar to spinel which is octohedral. Determination of RI with differentiate garnets.

Occurrence: Predominantly found in metamorphic rocks, especially contact metamorphic zones

Physical Properties

Name: Grunerite

Commodity: Asbestos

Formula: $(Mg,Fe)_7Si_8O_{22}(OH)_2$

Crystal System: Orthorhombic

Color: White, grey, greenish, brownish green, clove-brown, dark brown

Opacity: Transparent to nearly opaque

Luster: Vitreous to silky

Streak: Colorless or greyish

SGLow: 2.85

SGHigh: 3.57

HardnessLow: 5.5

HardnessHigh: 6

Cleavage: 3

Direction: {110} perfect, {010} and {100} imperfect

Habit: Fibrous to asbestiform

Fracture:

Other: Heat resistant, used for brake pads and fire retardent materials, roofing and cladding and fire retardent paints

Comments: Fibrous anthophyllite that occurs in metamorphic rocks

Optical Properties

Name: Grunerite

Group: Amphibole

Formula: $(Fe,Mg)_7Si_8O_{22}(OH)_2$

Crystal System: Monoclinic

Color: Neutral

Form: Fibrous to columnar aggregates, sometimes asbestiform. Cross sections are rhombic

Relief: High, n>balsam

Birefringence: Strong, 0.042-0.054. Sections with parallel extinction are first order.

2V: 79"-86"

Nalpha or Nord.: 1.657-1.663

NBeta or Nextr.: 1.684-1.697

NGamma: 1.699-1.717

Optical Sign: Biaxial negative

Orientation: Length slow, r>v weak

Pleochroism:

Twinning: Polysynthetic twinning characteristic {100}

Cleavage: In two directions at 56" and 124".

Extinction: Oblique 10"-15" in longitudinal sections

Alteration:

Features: Series with magnesiocummingtonite and cummingtonite. Smaller extiction angle than cummingtonite. RI higher than tremolite. Anthophyllite has parallel extinction.

Occurrence: Chiefly within metamorphic rocks, particularly banded iron formations and mica schists

Physical Properties

Name: Gypsum

Commodity: Gypsum

Formula: $CaSO_4+2H_2O$

Crystal System: Monoclinic

Color: Colorless, white, grey, yellowish, greenish, reddish, brownish

Opacity: Transparent

Luster: Subvitreous, pearly on cleavage

Streak: White

SGLow: 2.32

SGHigh: 2.32

HardnessLow: 2

HardnessHigh: 2

Cleavage: 3

Direction: {010} perfect, {100} and {011} distinct

Habit: Crystals thin to thick tabular, diamond-shaped, short to long prismatic; acicular; massive, granular, concretionary

Fracture: Splintery; flexible, not elastic

Other: Used as fertilizer, in cemet manufacture, as a filler in paper, rubber etc. and as plaster

Comments: Associated with anhydrite and halite in evaporite basins in arid climates

Optical Properties

Name: Gypsum

Group: Evaporite

Formula: $CaSO_4+2H_2O$

Crystal System: Monoclinic

Color: Colorless

Form: Anhedral to subhedral aggregates, fibrous

Relief: Low, n>balsam

Birefringence: Weak, 0.009

2V: 58"

Nalpha or Nord.: 1.520

NBeta or Nextr.: 1.522

NGamma: 1.529

Optical Sign: Biaxial positive

Orientation: Cleavage parallel to slow and fast rays

Pleochroism:

Twinning: Polysynthetic

Cleavage: Three perfect {010}, imperfect {100} and {1_11}

Extinction: Parallel to cleavage

Alteration:

Features: Sometimes fluoresces and phosphoresces greenish white. Weak birefringence and low relief compared to anhydrite.

Occurrence: The main constituent of gypsum rock, formed by hydration of anhydrite

Physical Properties

Name: Hafnium

Commodity: Hafnium (Ha)

Formula:

Crystal System:

Color:

Opacity:

Luster:

Streak:

SGLow:

SGHigh:

HardnessLow:

HardnessHigh:

Cleavage:

Direction:

Habit:

Fracture:

Other: Used in flashbulbs, nuclear reactors, enamels and refractory alloys

Comments: Obtained as a byproduct during production of zirconium sponge

Physical Properties

Name: Halite

Commodity: Rock Salt

Formula: NaCl

Crystal System: Cubic

Color: Colorless, white,yellow, orange, reddish, purple, blue

Opacity: Transparent to translucent

Luster: Vitreous

Streak:

SGLow: 2.17

SGHigh: 2.17

HardnessLow: 2

HardnessHigh: 2

Cleavage: 1

Direction: {001} perfect

Habit: Crystals cubic, rarely octahedral, often hopper-shaped or cavernous; massive, compact or granular; columnar; stalactitic

Fracture: Conchoidal; brittle

Other: Used in foods and chemical manufacturing

Comments: Occurs in arid climates where it forms deposits in evaporite lakes

Optical Properties

Name: Halite

Group: Evaporite

Formula: NaCl

Crystal System: Isometric, cubic

Color: Colorless

Form: Usually anhedral crystals

Relief: Very low, about the same as balsam

Birefringence: Nil, dark between crossed polars

2V:

Nalpha or Nord.: 1.544

NBeta or Nextr.:

NGamma:

Optical Sign:

Orientation:

Pleochroism:

Twinning:

Cleavage: Perfect, three directions

Extinction:

Alteration: Soluble in water

Features: Low relief, cubic form and solubility

Occurrence: Characteristic of sedimentary evaporite deposits

Physical Properties

Name: Hematite

Commodity: Iron (Fe)

Formula: Fe2O3

Crystal System: Trigonal

Color: Steel-grey to iron-black, thin fragments deep blood red

Opacity: Opaque

Luster: Metallic, submetallic, dull

Streak: Deep red or brownish red

SGLow: 5.26

SGHigh: 5.26

HardnessLow: 5

HardnessHigh: 6

Cleavage: 2

Direction: {0001} and {10-11} parting due to twinning

Habit: Crystals thin to thick tabular, rhombohedral, pyramidal, or prismatic; tabular crystals may form rosettes; massive

Fracture: Subconchoidal to uneven; brittle

Other: Iron and Steel

Comments: Hematite occurs in Archean Banded Iron Formations (BIF)

Optical Properties

Name: Hematite

Commodity: Iron (Fe)

Formula: Fe2O3

Crystal System: Trigonal

Color: Gray white with bluish tint

Pleochroism: Weak

Anisotropism: Distinct, Gray blue to Gray yellow

Internal Reflections: Deep red common

Reflectance 546nm: 26.1-30.2

Reflectance at 589nm: 25.1-29.15

Quant. Color Coordinates (QC): 0.299, 0.309, 29.8

Vickers Hardness Number (100g): 1038

Polishing Hardness: Greater than magnetite, less than pyrite

Cleavage: {0001} and {10-11} parting due to twinning

Form: Crystals thin to thick tabular, rhombohedral, pyramidal, or prismatic; tabulaBladed and acicular sub parallel and radiating aggregates common. Lamellar twinng frequent. Exsolution lenses or lamellae in ilmenite and magnetite or reverse often observed

Alteration: Exsolution lenses or lamellae in ilmenite and magnetite or reverse often observed

Associated Minerals: With magnetite, ilmenite, pyrite, chalcopyrite, bornite, rutile, cassiterite, sphalerite

Distinguishing Features: Deep red internal reflections and anisotropy charcteristic

Model: Hematite occurs in Archean Banded Iron Formations (BIF)

Physical Properties

Name: Hemimorphite

Commodity: Zinc (Zn)

Formula: Zn4Si2O7(OH)2+H2O

Crystal System: Orthorhombic

Color: White, colorless, pale blue, greenish, grey, yellowish, brown

Opacity: Transparent to translucent

Luster: Vitreous

Streak: Colorless

SGLow: 3.4

SGHigh: 3.5

HardnessLow: 4.5

HardnessHigh: 5

Cleavage: 3

Direction: {110} perfect, {101} imperfect, {001} traces

Habit: Crystals thin tabular, vertically striated, doubly terminated crystals distinctly hemimorphic, often in fan-shaped groups

Fracture: Subconchoidal to uneven; brittle

Other: Alloyed to make brass and zinc die castings. Corrosion resistant coatings on steel (galvanizing) and iron. Manufacture of corrosion resitant paints, pigments and fillers etc.

Comments: Occurs with other zinc ores in veins and replacement deposits

Optical Properties

Name: Hemimorphite

Commodity: Zinc (Zn)

Formula: Zn4Si2O7(OH)2+H2O

Crystal System: Orthorhombic

Color: White, colorless, pale blue, greenish, grey, yellowish, brown

Pleochroism: Transparent to translucent

Anisotropism: Vitreous

Internal Reflections: Colorless

Reflectance 546nm: 4.5

Reflectance at 589nm: 5

Quant. Color Coordinates (QC): 3.4

Vickers Hardness Number (100g): 3.5

Polishing Hardness: 3

Cleavage: {110} perfect, {101} imperfect, {001} traces

Form: Crystals thin tabular, vertically striated, doubly terminated crystals distinctly hemimorphic, often in fan-shaped groups

Alteration:

Associated Minerals:

Distinguishing Features: Crystals thin tabular, vertically striated, doubly terminated crystals distinctly hemimorphic, often in fan-shaped groups

Model: Occurs with other zinc ores in veins and replacement deposits

Physical Properties

Name: Ilmenite

Commodity: Titanium (Ti)

Formula: $FeTiO_3$

Crystal System: Trigonal

Color: Iron-black

Opacity: Opaque

Luster: Metallic to dull

Streak: Black

SGLow: 4.72

SGHigh: 4.72

HardnessLow: 5

HardnessHigh: 6

Cleavage: 2

Direction: {0001} and {10-11} parting

Habit: Crystals thick tabular or acute rhombohedral; granular; massive, lamellar to compact

Fracture: Conchoidal to uneven; brittle

Other: High strength and corrosion resistance. Mostly used in aircraft and aerospace industries. Used in chemical and desalination plants and heatbexchanger tubing in power plants

Comments: Associated with monazite, zircon, rutile etc in alluvial sands and heavy mineral strandline beach deposits. An accessory mineral in igneous rocks concentrated by sedimentary processes

Optical Properties

Name: Ilmenite

Commodity: Titanium (Ti) and Ilmenite

Formula: $FeTiO_3$

Crystal System: Trigonal, hexagonal

Color: Brownish with pinkish or violet tint

Pleochroism: Distinct, pinkish brown to dark brown

Anisotropism: Strong, greenish Gray to brownish Gray

Internal Reflections: Rare, dark brown

Reflectance 546nm: 17.0-20.1

Reflectance at 589nm: 17.4-20.2

Quant. Color Coordinates (QC):

Vickers Hardness Number (100g): 659-703, convex

Polishing Hardness: Greater than magnetite, less than hematite

Cleavage: {0001} and {10-11} parting

Form: Anhedral to subhedral crystals and exsolution lamellae or lenses in hematite and magnetite. Lamelllar twinning frequent

Alteration: Exsolution lamellae or lenses in hematite and magnetite

Associated Minerals: Widespread accessory in igneous and metamorphic rocks, with magnetite, hematite, rutile, pyrite, pyrrhotite, chromite, pentlandite, tantalite

Distinguishing Features: Crystals thick tabular or acute rhombohedral; granular; massive, lamellar to compact. Anisotropy and pleochroism distinctive

Model: Associated with monazite, zircon, rutile etc in alluvial sands and heavy mineral strandline beach deposits. An accessory mineral in igneous rocks concentrated by sedimentary processes

Physical Properties

Name: Indium

Commodity: Indium (In)

Formula:

Crystal System:

Color:

Opacity:

Luster:

Streak:

SGLow:

SGHigh:

HardnessLow:

HardnessHigh:

Cleavage:

Direction:

Habit:

Fracture:

Other: Alloyed for low melting point metals and solders

Comments: No ores. Recovered from flue dusts and residue in some zinc smelters

Optical Properties

Name: Indium

Commodity: Indium (In)

Formula:

Crystal System:

Color:

Pleochroism:

Anisotropism:

Internal Reflections:

Reflectance 546nm:

Reflectance at 589nm:

Quant. Color Coordinates (QC):

Vickers Hardness Number (100g):

Polishing Hardness:

Cleavage:

Form:

Alteration:

Associated Minerals:

Distinguishing Features:

Model: No ores. Recovered from flue dusts and residue in some zinc smelters

Physical Properties

Name: Jamesonite

Commodity: Lead (Pb)

Formula: Pb4FeSb6S14

Crystal System: Monoclinic

Color: Grey-black, may tarnish iridescent

Opacity: Opaque

Luster: Metallic

Streak: Grey-black

SGLow: 5.63

SGHigh: 5.63

HardnessLow: 2.5

HardnessHigh: 2.5

Cleavage: 1

Direction: {001} good

Habit: Crystals acicular to fibrous; felt-like aggregates; columnar aggregates; massive

Fracture: Brittle

Other: Pipes, alloys, batteries, radiation shielding, pigments

Comments: Occurs in veins with galena, sphalerite, pyrite and stibnite

Optical Properties

Name: Jamesonite

Commodity: Lead (Pb)

Formula: Pb4FeSb6S14

Crystal System: Monoclinic

Color: White

Pleochroism: Strong, white to yellow green

Anisotropism: Strong, Gray, tan, brown, blue

Internal Reflections: Reddish in Bi rich variety

Reflectance 546nm: 37.4-42.9

Reflectance at 589nm: 36.7-41.5

Quant. Color Coordinates (QC):

Vickers Hardness Number (100g): 113-117, perfect

Polishing Hardness: Less than galena

Cleavage: Parallel to longitudinal sections common

Form: Acicular and lathlike aggregates, twinning frequent

Alteration:

Associated Minerals: With galena, pyrite, pyargyrite, boulangerite, chalcopyrite, sphalerite, tetrahedrite, arsenopyrite

Distinguishing Features: Strong pleochroism and anisotropy

Model: Occurs in veins with galena, sphalerite, pyrite and stibnite

Physical Properties

Name: Kaolinite
Commodity: China Clay
Formula: $Al_2Si_2O_5(OH)_4$
Crystal System: Triclinic
Color: Colorless, white, yellowish pink, reddish, bluish
Opacity: Transparent to translucent
Luster: Pearly to dull earthy
Streak:
SGLow: 2.6
SGHigh: 2.63
HardnessLow: 2
HardnessHigh: 2.5
Cleavage: 1
Direction: {001} perfect
Habit: Thin hexagonal platelets or scales; elongated platelets or curved laths; massive compact, friable, mealy
Fracture: Scales flexible, inelastic, often plastic when moist
Other: Manufactur of porcelain, china and bricks. Used as filler in manufacture of paper, rubber and paint
Comments: Occurs in the duricrust in wet climates from weathering of aluminous silicates such as K-feldspar

Optical Properties

Name: Kaolinite
Group: Kaolinite, Clay
Formula: $Al_2Si_2O_5(OH)_4$
Crystal System: Triclinic
Color: Colorless to pale yellow
Form: Microcrystalline aggregates of scales and plates with six sided outline, in veins replacing feldspars
Relief: Low, n>balsam
Birefringence: Weak, 0.005
2V: Variable
Nalpha or Nord.: 1.561
NBeta or Nextr.: 1.565
NGamma: 1.566
Optical Sign: Biaxial negative
Orientation: Length slow
Pleochroism:
Twinning:
Cleavage: One perfect {001}
Extinction: Almost parallel 1'-3.5" along base
Alteration:
Features: Low birefringence, aggregates of hexagonal scales and plates
Occurrence: A weathering product of igneous and metamorphic rocks and in clay beds. Also common as a diagenetic clay mineral in sedimentary beds

Physical Properties

Name: Kernite

Commodity: Borates

Formula: Na2B4O6(OH)2+3H2O

Crystal System: Monoclinic

Color: Colorless; white when surface altered

Opacity: Transparent

Luster: Dull to vitreous, silky or pearly on cleavage

Streak:

SGLow: 1.91

SGHigh: 1.91

HardnessLow: 2.5

HardnessHigh: 3

Cleavage: 3

Direction: {100} and {001} perfect; {-201} distinct

Habit: Crystals nearly equant; massive, fibrous structure, cleavable

Fracture: Brittle, splintery

Other: Used in manufacture of insulating fibreglass, as fluxes for manufacture of glass and enamels. Borax is used in cloth manufacture and tanning, soap and glue, as preservatives and antiseptics

Comments: Important source of borates. Precipitated in evaporite basins in arid climates

Optical Properties

Name: Kernite

Group: Borate

Formula: Na2B4O6(OH)2+3H2O

Crystal System: Monoclinic

Color: Colorless

Form: Large crystals upto 1m thick

Relief: Low to moderate, n< balsam

Birefringence: Moderate to strong, 0.034

2V: 80"

Nalpha or Nord.: 1.454

NBeta or Nextr.: 1.472

NGamma: 1.488

Optical Sign: Biaxial negative

Orientation:

Pleochroism:

Twinning:

Cleavage: Three {100} perfect, {001} good, {201} fair

Extinction:

Alteration:

Features: Similar to gypsum, but soluble in warm water

Occurrence: Occurs within beds in desert environments

Physical Properties
Name: Kyanite
Commodity: Aluminum Silicate
Formula: Al2SiO5
Crystal System: Triclinic
Color: Blue, colorless, white, grey, green, yellow, pink, nearly black
Opacity: Transparent to translucent
Luster: Vitreous to pearly
Streak: Colorless
SGLow: 3.53
SGHigh: 3.67
HardnessLow: 4
HardnessHigh: 7.5
Cleavage: 3
Direction: {100} perfect, {010} distinct, {001} parting
Habit: Crystals long bladed, elongated parallel to c-axis; often bent or twisted; massive, bladed to fibrous
Fracture: Somewhat flexible
Other: High alumina refractory for iron and steel industries and other metal smelters. Used in glass industry and as insulating porcelain
Comments: Widely distributed in schists, gneisses, pegmatites and quartz veins in metamorphic belts

Optical Properties
Name: Kyanite
Group: Sillimanite
Formula: Al2SiO5
Crystal System: Triclinic
Color: Colorless to pale blue
Form: Broad, tabular plates parallel to {100} and narrow sections parallel to {010}. Often bent crystals
Relief: High, n>balsam
Birefringence: Moderate, 0.016
2V: 82"
Nalpha or Nord.: 1.712
NBeta or Nextr.: 1.720
NGamma: 1.728
Optical Sign: Biaxial negative
Orientation: Length slow, r>v
Pleochroism: May be p[leochroic in thick sections
Twinning: Frequent along {100} and {001}
Cleavage: Three perfect parallel to {100}, distinct parallel to {010} at right angles. Parting at 85" to length.
Extinction: Oblique 30" in longitudinal sections, parallel or nearly so in cross sections
Alteration:
Features: Extinction angle of 30" and biaxial figure are distinctive
Occurrence: Widespread within metamorphic rocks, including schists, gneisses, some eclogites and pegmatites.

Physical Properties

Name: Lepidolite

Commodity: Lithium (Li) , Rubidium (Ru), Cesium (Ce)

Formula: K(Li,Al)3(Si,Al)4O10(F,OH)2

Crystal System: Monoclinic

Color: Pink to purple, colorless, white, greyish, yellowish

Opacity: Transparent to translucent

Luster: Pearly

Streak: Colorless

SGLow: 2.8

SGHigh: 3.3

HardnessLow: 2.5

HardnessHigh: 3

Cleavage: 3

Direction: {001} perfect, micaceous, {110} and {010} imperfect

Habit: Crystals tabular, pseudohexagonal or hexagonal; thick cleavable masses, coarse to fine scaly; aggregates of books

Fracture: Flexible, inelastic

Other: Used as a base for greases, in production of aluminum and in ceramics

Comments: Asociated with tourmaline and spodumene. Occurs in pegamtites and schists

Optical Properties

Name: Lepidolite

Group: Mica

Formula: K(Li,Al)3(Si,Al)4O10(F,OH)2

Crystal System: Monoclinic

Color: Colorless

Form: Short prismatic pseudo-hexagonal and thick tabular crystals

Relief: Fair, n>balsam

Birefringence: Strong, 0.045

2V: 40"

Nalpha or Nord.: 1.560

NBeta or Nextr.: 1.598

NGamma: 1.605

Optical Sign: Biaxial negative

Orientation: Length slow, r>v weak

Pleochroism:

Twinning: according to mica law {110}

Cleavage: One perfect {001}

Extinction: Oblique to parallel 0"-7" along cleavage

Alteration:

Features: Similar to muscovite but larger axial angle. Occurence with other lithium minerals may be distinctive

Occurrence: A lithium mineral found mainly in granite pegmatites with other lithium minerals, spodumene

Physical Properties
Name: Limonite
Group: Limonite-Goethite
Formula: FeO.OH.nH2O
Crystal System: Amorphous/ Cryptocrystalline
Color: Yellow, brown, brownish-black, orange-brown
Opacity: Opaque
Luster: Non metallic
Streak: Yellowish-brown
SGLow: 2.7
SGHigh: 4.3
HardnessLow: 4
HardnessHigh: 5.5
Cleavage: None
Direction: None
Habit: Massive, earthy, vitreous
Fracture: Uneven, subconchoidal
Other: Secondary iron oxide, usually the product of weathering
Comments: A general term for hydrous iron oxides

Optical Properties
Name: Limonite
Group: Limonite-Goethite
Formula: FeO.OH.nH2O
Crystal System: Amorphous/ Cryptocrystalline
Color: Yellowish-brown
Form: Massive, earthy, vitreous
Relief:
Birefringence:
2V:
Nalpha or Nord.:
NBeta or Nextr.:
NGamma:
Optical Sign:
Orientation:
Pleochroism:
Twinning:
Cleavage:
Extinction:
Alteration:
Features: A general term for hydrous iron oxides
Occurrence: Secondary iron oxide, usually the product of weathering

Physical Properties

Name: Linnaeite

Commodity: Cobalt (Co)

Formula: Co3S4

Crystal System: Cubic

Color: Light grey to steel-grey, rapidly tarnishes copper-red or violet-grey

Opacity: Opaque

Luster: Metallic

Streak:

SGLow: 4.8

SGHigh: 5.0

HardnessLow: 4.5

HardnessHigh: 5.5

Cleavage: 1

Direction: {001} imperfect

Habit: Crystals octahedral; massive, compact to granular

Fracture: Subconchoidal to uneven

Other: Alloying to produce high temperature steels and magnetic alloys. Used as a catalyst in the chemical industries and as cement in sintered carbide cutting equipment

Comments: Sometimes occurs in small amounts in copper ores. Cobalt is extracted as a byproduc during lead, copper and nickel smelting

Optical Properties

Name: Linnaeite

Commodity: Cobalt (Co)

Formula: Co3S4

Crystal System: Cubic

Color: Creamy white

Pleochroism: None

Anisotropism: Isotropic

Internal Reflections: None

Reflectance 546nm: 49.5

Reflectance at 589nm: 49.6

Quant. Color Coordinates (QC):

Vickers Hardness Number (100g): 450-613

Polishing Hardness: Greater than chlacopyrite, sphalerite, less than arsenopyrite

Cleavage:

Form: Euhedral and subhedral crystal aggregates. Lamellar intergrowths with millerite, chalcopyrite, bornite, pyrrhotite, pyrite, bismuth, covellite, safflorite, niccolite

Alteration: Lamellar intergrowths with millerite, chalcopyrite, bornite, pyrrhotite, pyrite, bismuth, covellite, safflorite, niccolite

Associated Minerals: Lamellar intergrowths with millerite, chalcopyrite, bornite, pyrrhotite,, pyrite, bismuth, covellite, safflorite, niccolite

Distinguishing Features: Cubic and octohedral crystals, isotropism

Model: Sometimes occurs in small amounts in copper ores. Cobalt is extracted as a byproduc during lead, copper and nickel smelting

Optical Properties

Name: Mackinawite

Commodity: Iron (Fe)

Formula: Fe1+xS

Crystal System: Tetragonal

Color: Pinkish to redish Gray

Pleochroism: Moderate to strong, pinkish Gray to Gray

Anisotropism: Very strong, Grayish white to bluish to brownish

Internal Reflections: None

Reflectance 546nm: 22-46

Reflectance at 589nm:

Quant. Color Coordinates (QC):

Vickers Hardness Number (100g): 52-58

Polishing Hardness: Equal to pyrrhotite

Cleavage:

Form: Wormlike grains and lamellae in pyrite, chalcopyrite, cubanite and pentlandite

Alteration:

Associated Minerals: Wormlike grains and lamellae in pyrite, chalcopyrite, cubanite and pentlandite

Distinguishing Features: Appears as bright grains when polars are nearly crossed.

Optical Properties

Name: Maghemite

Commodity: Iron (Fe)

Formula: Fe3O4

Crystal System: Cubic

Color: Bluish Gray

Pleochroism: None

Anisotropism: Isotropic

Internal Reflections: Rare, brownish red

Reflectance 546nm: 24.4

Reflectance at 589nm: 28.8

Quant. Color Coordinates (QC): 0.293, 0.304, 24.1

Vickers Hardness Number (100g): 412 (50g)

Polishing Hardness: Greater than magnetite, less than hematite

Cleavage:

Form: As lamellae and patches in magnetite

Alteration: Oxidation product of magnetite, rare

Associated Minerals: As lamellae and patches in magnetite

Distinguishing Features: As lamellae and patches in magnetite

Physical Properties

Name: Magnesite

Commodity: Magnesium (Mg)

Formula: $MgCO_3$

Crystal System: Trigonal

Color: Colorless, white, grey, yellowish to brown

Opacity: Transparent to translucent

Luster: Vitreous, dull

Streak:

SGLow: 3.0

SGHigh: 3.1

HardnessLow: 3.75

HardnessHigh: 4.25

Cleavage: 1

Direction: {10-11} perfect

Habit: Crystals rhombohedral, uncommon; rarely prismatic, tabular or scalenohedral; massive, compact, chalky, lamellar, fibrous

Fracture: Conchoidal; brittle

Other: Alloyed with aluminum to im,prove strength and corrosion resistance. Manufacture of Iron nodules and desulphuring. Used in petrol, zirconium and titanium production and magnesium die-castings

Comments: Associated with serpentinite. Occurs as an alteration product of olivine and pyroxenes in metamorphic belts, particularly Archean greenstone belts

Optical Properties

Name: Magnesite

Group: Calcite

Formula: $MgCO_3$

Crystal System: Hexagonal

Color: Colorless

Form: Anhedral to subhedral crystal aggregates. Euhedral crystals rare.

Relief: High along long direction, low along short direction

Birefringence: Extreme, 0.191-0.199

2V:

Nalpha or Nord.: 1.509-1.527

NBeta or Nextr.: 1.700-1.726

NGamma:

Optical Sign: Uniaxial negative

Orientation:

Pleochroism:

Twinning: None

Cleavage: One perfect rhombohedral along {101_1}

Extinction: Symmetrical to cleavage

Alteration:

Features: Similar to dlomite and calcite. Staining may be required.

Occurrence: Main constituent of metamorphic magnesite rocks. Also occurs in serpentinites

Physical Properties
Name: Magnetite
Commodity: Iron (Fe)
Formula: Fe3O4
Crystal System: Cubic
Color: Iron-black, greyish black
Opacity: Opaque
Luster: Splendent metallic to dull
Streak: Black
SGLow: 5.175
SGHigh: 5.175
HardnessLow: 5.5
HardnessHigh: 6.5
Cleavage: 1
Direction: {111} parting good
Habit: Crystals octahedral, may be highly modified or dodecahedral and striated; massive, compact or fine to coarse granular
Fracture: Subconchoidal to uneven; brittle
Other: Iron and Steel
Comments: Occurs in plutonic and metamorphic rocks and Archean Banded Iron Formations (BIF)

Optical Properties
Name: Magnetite
Commodity: Iron (Fe)
Formula: Fe3O4
Crystal System: Cubic
Color: Gray with brownish tint
Pleochroism: None
Anisotropism: Isotropic, weak anomolous anisotropy
Internal Reflections: None
Reflectance 546nm: 20.0
Reflectance at 589nm: 20.3
Quant. Color Coordinates (QC): 0.311, 0.314, 20.1
Vickers Hardness Number (100g): 592, perfect
Polishing Hardness: Greater than pyrhhotite, less than ilmenite, hematite, pyrite
Cleavage: {111} parting good
Form: Subhedral to euhedral prismatic crystals and anhedral polycrystalline aggregates. Also skeletal crystals
Alteration: Exsolution or oxidation lamellae of hematite, ilmenite and ulvospinel. Alters to hematite and goethite
Associated Minerals: With Pyrrhotite, pyrite, pentlandite, chalcopyrite, bornite, sphalerite and galena
Distinguishing Features: Crystals octahedral, may be highly modified or dodecahedral and striated; massive, compact or fine to coarse granular
Model: Occurs in plutonic and metamorphic rocks and Archean Banded Iron Formations (BIF)

Physical Properties
Name: Malachite
Commodity: Copper (Cu)
Formula: Cu2(CO3)(OH)2
Crystal System: Monoclinic
Color: Bright green to dark or blackish green
Opacity: Translucent to opaque
Luster: Vitreous to adamantine
Streak: Pale green
SGLow: 4.05
SGHigh: 4.05
HardnessLow: 3.5
HardnessHigh: 4
Cleavage: 2
Direction: {201} perfect, {010} fair
Habit: Crystals acicular or short to long prismatic, wedge-shaped terminations, small; massive; compact crusts; botryoidal
Fracture: Subconchoidal to uneven; crystals brittle; massive material tough
Other: Alloyed with various metals to produce bronzes and brasses. Used in electrical industries where high electrical and thermal conductivity is required.
Comments: Chalcopyrite is the main ore of copper. Malachite is found in the oxidized zone of copper deposits

Optical Properties
Name: Malachite
Commodity: Copper (Cu)
Formula: Cu2(CO3)(OH)2
Crystal System: Monoclinic
Color: Gray, masked by internal reflections
Pleochroism:
Anisotropism:
Internal Reflections: Masked by internal reflections
Reflectance 546nm:
Reflectance at 589nm:
Quant. Color Coordinates (QC):
Vickers Hardness Number (100g):
Polishing Hardness:
Cleavage: {201} perfect, {010} fair
Form: Crystals acicular or short to long prismatic, wedge-shaped terminations, small; massive; compact crusts; botryoidal
Alteration:
Associated Minerals:
Distinguishing Features:
Model: Chalcopyrite is the main ore of copper. Malachite is found in the oxidized zone of copper deposits

Physical Properties

Name: Manganite

Commodity: Manganese (Mn)

Formula: MnO(OH)

Crystal System: Monoclinic

Color: Black to dark steel-grey

Opacity: Opaque

Luster: Submetallic to dull

Streak: Reddish brown to black

SGLow: 4.33

SGHigh: 4.33

HardnessLow: 4

HardnessHigh: 4

Cleavage: 3

Direction: {010} perfect, {110} and {001} imperfect

Habit: Crystals short to long prismatic, vertically striated, often grouped in bundles; massive, fibrous to columnar

Fracture: Uneven; brittle

Other: Important iron alloy for steel making

Comments: Associated with baryte, pyrolusite and goethite in veins in granite

Optical Properties

Name: Manganite

Commodity: Manganese (Mn)

Formula: MnO(OH)

Crystal System: Monoclinic

Color: Gray to brownish Gray

Pleochroism: Weak, brownish Gray

Anisotropism: Isotropic

Internal Reflections: None

Reflectance 546nm: 14.8-20.7

Reflectance at 589nm: 14.3-19.9

Quant. Color Coordinates (QC):

Vickers Hardness Number (100g): 698-772, perfect

Polishing Hardness: Less than haussmanite, jacobsite

Cleavage: May be visible on {010} and {110}

Form: Prismatic and lamellar crystal aggregates. Frequently intergrown with pyrolusite and psilomelane

Alteration: Frequently intergrown with pyrolusite and psilomelane

Associated Minerals: Pyrolusite, psilomelane, hausmannite, braunite, goethite

Distinguishing Features: Prismatic crystals with pyrolusite and psilomelane

Physical Properties

Name: Manganotantalite

Commodity: Tantalum (Ta)

Formula: MnTa2O6

Crystal System: Orthorhombic

Color: Brownish black with reddish brown or red internal reflections

Opacity: Translucent

Luster: Vitreous to resinous

Streak: Dark red

SGLow: 8.003

SGHigh: 8.003

HardnessLow: 6

HardnessHigh: 6.5

Cleavage: 2

Direction: {010} distinct, {100} indistinct

Habit: Crystals short prismatic, equant or thick tabular; massive

Fracture: Subconchoidal to uneven; brittle

Other: Very high corrosion resistance. Used in production of special steels for medical use, in chemical and electrical processes, electrodes, cutting tools and manufacture of electrodes

Comments: Associated with cassiterite, wolframite, spodumene and tourmaline in granite pegmatites. Tantalum is also recovered from tin slags as a byproduct

Optical Properties

Name: Manganotantalite

Commodity: Tantalum (Ta)

Formula: MnTa2O6

Crystal System: Orthorhombic

Color:

Pleochroism:

Anisotropism:

Internal Reflections:

Reflectance 546nm:

Reflectance at 589nm:

Quant. Color Coordinates (QC):

Vickers Hardness Number (100g):

Polishing Hardness:

Cleavage: {010} distinct, {100} indistinct

Form: Crystals short prismatic, equant or thick tabular; massive

Alteration:

Associated Minerals:

Distinguishing Features: Crystals short prismatic, equant or thick tabular; massive

Model: Associated with cassiterite, wolframite, spodumene and tourmaline in granite pegmatites. Tantalum is also recovered from tin slags as a byproduct

Physical Properties

Name: Marcasite

Commodity:

Formula: FeS2

Crystal System: Orthorhombic

Color: Pale brass-yellow to tin-white, darkens on exposure

Opacity: Opaque

Luster: Metallic

Streak: Greenish black

SGLow: 4.92

SGHigh: 4.92

HardnessLow: 6

HardnessHigh: 6.5

Cleavage: 2

Direction: {101} distinct; {110} traces

Habit: Crystals tabular or pyramidal, prismatic or capillary; massive, fine granular; stalactitic; globular; reniform

Fracture: Uneven; brittle

Other:

Comments: Usually regarded as a gangue mineral in sulphide ore deposits. Occurs in hydrothermal veins and replacement deposits

Optical Properties

Name: Marcasite

Commodity:

Formula: FeS2

Crystal System: Orthorhombic

Color: Yellowish white with slight pinkish or greenish tint

Pleochroism: Strong, brownish, yellowish green

Anisotropism: Strong, blue to green yellow to purple Gray

Internal Reflections: None

Reflectance 546nm: 48.2-55.8

Reflectance at 589nm: 48.4-54.6

Quant. Color Coordinates (QC): 4.92

Vickers Hardness Number (100g): 1288-1681, fractured

Polishing Hardness: Equal to pyrite

Cleavage: {101} distinct; {110} traces

Form: Subhedral to lamellar intergrowths with pyrite, vienlets in pyrhhotite and iron oxides, radiating colloform bands, frequently twinned

Alteration:

Associated Minerals: Usually with pyrite and with other sulphides

Distinguishing Features: Blue to yellowish anisotropy distinctive

Model: Usually regarded as a gangue mineral in sulphide ore deposits. Occurs in hydrothermal veins and replacement deposits

Physical Properties

Name: Marmatite (Fe bearing Sphalerite)

Commodity: Zinc (Zn)

Formula:

Crystal System:

Color:

Opacity:

Luster:

Streak:

SGLow:

SGHigh:

HardnessLow:

HardnessHigh:

Cleavage:

Direction:

Habit:

Fracture:

Other: Alloyed to make brass and zinc die castings. Corrosion resistant coatings on steel (galvanizing) and iron. Manufacture of corrosion resitant paints, pigments and fillers etc.

Comments: Usually associated with galena in zinc ore deposits

Optical Properties

Name: Marmatite (Fe bearing Sphalerite)

Commodity: Zinc (Zn)

Formula:

Crystal System:

Color:

Pleochroism:

Anisotropism:

Internal Reflections:

Reflectance 546nm:

Reflectance at 589nm:

Quant. Color Coordinates (QC):

Vickers Hardness Number (100g):

Polishing Hardness:

Cleavage:

Form:

Alteration:

Associated Minerals:

Distinguishing Features:

Model: Usually associated with galena in zinc ore deposits

Optical Properties

Name: Maucherite

Commodity: Nickel (Ni)

Formula: Ni11As8

Crystal System: Tetragonal

Color: White

Pleochroism: None

Anisotropism: Weak to distinct in oil

Internal Reflections: None

Reflectance 546nm: 47.8-48.5

Reflectance at 589nm: 50.0-50.7

Quant. Color Coordinates (QC):

Vickers Hardness Number (100g): 715-743, perfect

Polishing Hardness: Greater than chalcopyrite, sphalerite, less than saffloriteloellingite

Cleavage:

Form: Euhedral prismatic crystals and anhedral aggregates, sometimes twinned. Intergrowths with niccolite and gersdorffite

Alteration:

Associated Minerals: With chalcopyrite, cubanite, siegenite

Distinguishing Features: White color, prismatic crystals, occurrence

Optical Properties

Name: Mawsonite

Commodity: Copper (Cu)

Formula: $Cu_7Fe_2SnS_{10}$

Crystal System: Tetragonal

Color: Brownish orange

Pleochroism: Strong, orange to brown

Anisotropism: Very strong, straw yellow to royal blue

Internal Reflections: None

Reflectance 546nm: 26.9-29.7

Reflectance at 589nm: 29.1-35.1

Quant. Color Coordinates (QC): 0.339, 0.340, 27.3

Vickers Hardness Number (100g): 166-210

Polishing Hardness: Greater than bornite

Cleavage:

Form: Inclusions within bornite

Alteration:

Associated Minerals: Chalcopyrite, chalcocite, bornite, tetrahedrite, pyrite, galena, enargite, stannite

Distinguishing Features: Brownish orange color, inclusions

Physical Properties

Name: Microcline

Commodity: Feldspar

Formula: KAlSi3O8

Crystal System: Triclinic

Color: White, grey, yellowish, reddish, green

Opacity: Transparent to translucent

Luster: Vitreous, often pearly on cleavage

Streak:

SGLow: 2.55

SGHigh: 2.63

HardnessLow: 6

HardnessHigh: 6.5

Cleavage: 6

Direction: {001} and {010} perfect, {100}, {110}, {-110} and {-201} parting

Habit: Crystals short prismatic, blocky; or tabular; massive, cleavable to granular; twinning on Carlsbad, Manebach, Baveno laws

Fracture: Uneven; brittle

Other: Manufacture of porcelain, pottery and glass. Used for glazes in pottery and as mild abrasive

Comments: Important rock forming mineral widely distributed in igneous and metamorphic rocks. Also present in pegmatites

Optical Properties

Name: Microcline

Group: Feldspar

Formula: KAlSi3O8

Crystal System: Triclinic

Color: Colorless

Form: Subhedral to anhedral coarse to fine crystals. Intergrown with microcline to form perthite

Relief: Low, n<balsam

Birefringence: Weak, 0.007

2V: 77"-84"

Nalpha or Nord.: 1.518-1.522

NBeta or Nextr.: 1.522-1.526

NGamma: 1.525-1.530

Optical Sign: Biaxial negative

Orientation: Length fast, Dispersion r>v

Pleochroism:

Twinning: Polysynthetic twinning almost always occurs, according to albite law or pericline law, giving quadrille structure with twinning at right angles.

Cleavage: Three directions, perfect {001}, distinct {010}, imperfect {110} and {11_0}

Extinction: Oblique on {001} >15", on {010} >5"

Alteration: Alters to clay

Features: Polysynthetic twinning distinctive. Extinction angle of 15" greater than albite

Occurrence: Most abundant in felsic igneous rocks. Widespread occurrence

Physical Properties
Name: Millerite
Commodity: Nickel (Ni)
Formula: NiS
Crystal System: Trigonal
Color: Brass-yellow, bronze-yellow; tarnishes greenish grey
Opacity: Opaque
Luster: Metallic
Streak: Greenish black
SGLow: 5.41
SGHigh: 5.42
HardnessLow: 3
HardnessHigh: 3.5
Cleavage: 2
Direction: {01-12} and {10-11} perfect
Habit: Crystals slender acicular to capillary, elongated on c-axis; in radiating groups; fibrous masses; massive, cleavable
Fracture: Uneven; brittle
Other: Iron alloy for making stainless steel due to corrosion resistence, also non-ferrous alloys. Non toxic, used for food handling and pharmaceutical equipment. Used for electroplating steels and as a base for chromium plate
Comments: Associated with other nickel ores where it forms needle like radiating crystals replacing these ores

Optical Properties
Name: Millerite
Commodity: Nickel (Ni)
Formula: NiS
Crystal System: Trigonal, hexagonal
Color: Yellow
Pleochroism: Distinct in oil, yellow to blue or violet
Anisotropism: Strong, lemon yellow to blue or violet
Internal Reflections: None
Reflectance 546nm: 50.2-56.6
Reflectance at 589nm: 51.9-59.05
Quant. Color Coordinates (QC): 0.328, 0.339, 50.4
Vickers Hardness Number (100g): 196-376
Polishing Hardness: Greater than chalcopyrite, less than pentlandite
Cleavage: Usually visible on {1011}, {01-12} and {10-11} perfect
Form: Small plates and irregular inclusions. Rosettes and colloform bands infrequent. Twinning common
Alteration: Replaces other Ni-bearing sulphides
Associated Minerals: With Ni-bearing sulphides
Distinguishing Features: Color, anisotropy and occurence with nickel sulphides
Model: Associated with other nickel ores where it forms needle like radiating crystals replacing these ores

Physical Properties

Name: Molybdenite

Commodity: Molybdenum (Mo)

Formula: MoS2

Crystal System: Hexagonal

Color: Lead-grey

Opacity: Opaque

Luster: Metallic

Streak: Greenish

SGLow: 4.62

SGHigh: 5.06

HardnessLow: 1

HardnessHigh: 1.5

Cleavage: 1

Direction: {0001} perfect, easy

Habit: Crystals thin to thick tabular, hexagonal aspect; short barrel-shaped prisms; foliated or radiate masses; scales; grains

Fracture: Laminae flexible; sectile

Other: A ferro-alloy. Used in manufacture of electrodes and furnace parts and as a catalyst, corrosion inhibitor and additive to lubricants

Comments: Occurs in granites and pegmatites with cassiterite and wolfram

Optical Properties

Name: Molybdenite

Commodity: Molybdenum (Mo)

Formula: MoS2

Crystal System: Hexagonal

Color: White to Gray

Pleochroism: Extreme, white to Gray with bluish tint

Anisotropism: Very strong, white with pinkish tint, dark blue with nearly crossed polars

Internal Reflections: None

Reflectance 546nm: 19.5-38.5

Reflectance at 589nm: 19.0-38.8

Quant. Color Coordinates (QC): 0.298, 0.299, 39.3

Vickers Hardness Number (100g): 8-100, 32-33, parallel to cleavage

Polishing Hardness: less than most minerals

Cleavage: Usually visible{ 0001} perfect, undulatory extinction

Form: Usually small plates, often deformed, rosettes and colloform bands. Cleavage, twinning frequent.

Alteration:

Associated Minerals: In veins with pyrite, chalcopyrite, bornite, cassiterite, wolframite, bismuth, bismuthinite. Also with other sulphides

Distinguishing Features: Softness, pleochroism and anisotropy distinctive

Model: Occurs in granites and pegmatites with cassiterite and wolfram

Physical Properties

Name: Monazite-[Ce]

Commodity: Rare Earths (Ce, La, Y)

Formula: (Ce,La,Nd,Th)PO4

Crystal System: Monoclinic

Color: Reddish brown, brown, yellowish brown, pink, yellow, greenish, greyish white

Opacity: Transparent to subtranslucent

Luster: Resinous, waxy, or vitreous to subadamantine

Streak: White, slightly colored

SGLow: 4.6

SGHigh: 5.4

HardnessLow: 5

HardnessHigh: 5.5

Cleavage: 5

Direction: {100} distinct, {010} less distinct, {110}, {101} and {011} indistinct

Habit: Crystals thin to thick tabular, small, elongated on c-axis; also equant, wedge-shaped or prismatic; massive granular

Fracture: Conchoidal to uneven; brittle

Other: Cerium subgroup is most important of rare earths. Used as a catalyst in petroleum refining, iron-cerium alloys in lighter flints and used in ceramic and glass industries and for manufacture of color televisions

Comments: Occurs in granites and in heavy mineral strandline beach deposits

Optical Properties

Name: Monazite

Group: Monazite

Formula: (Ce,La,Nd,Th)PO4

Crystal System: Monoclinic

Color: Colorlesss to nuetral

Form: Small, euhedral crystals

Relief: V.high, n>balsam

Birefringence: Strong, 0.049-0.051. In cross sections weak, 0.001

2V: 6"-19"

Nalpha or Nord.: 1.786-1.800

NBeta or Nextr.: 1.788-1.801

NGamma: 1.837-1.849

Optical Sign: Biaxial positive

Orientation: Length slow

Pleochroism:

Twinning:

Cleavage: Five {100} distinct, {010} less distinct, {110}, {101} and {011} indistinct. Parting parallel to {001} often occurs

Extinction: Almost parallel 2" along length. Incomplete in sections parallel {001}

Alteration:

Features: Radioactive, euhedral crystals, birefringence

Occurrence: Occurs within pegmatites and granites and as a detrital mineral in sedimentary beds

Physical Properties

Name: Mullite

Commodity: Aluminum Silicate

Formula: Al6Si2O13

Crystal System: Orthorhombic

Color: Colorless to pale pink

Opacity: Transparent to translucent

Luster: Vitreous

Streak:

SGLow: 3.03

SGHigh: 3.16

HardnessLow: 6

HardnessHigh: 7

Cleavage: 1

Direction: {010} distinct

Habit: Crystals prismatic

Fracture:

Other: High alumina refractory for iron and steel industries and other metal smelters. Used in glass industry and as insulating porcelain

Comments: Mostly manufactured (synthetic)

Optical Properties

Name: Mullite

Group: Sillimanite

Formula: Al6Si2O13

Crystal System: Orthorhombic

Color: Colorless

Form: Long prismatic crystals with square cross sections

Relief: High, n>balsam

Birefringence: Weak, 0.012

2V: 20"

Nalpha or Nord.: 1.642

NBeta or Nextr.: 1.644

NGamma: 1.654

Optical Sign: Biaxial positive

Orientation: Length slow, r>v

Pleochroism:

Twinning:

Cleavage: One distinct {010}

Extinction: Parallel in longitudinal sections and symmetrical in cross sections

Alteration:

Features: Almost identical to sillimanite but RI is lower. Occurrence distinctive.

Occurrence: Rare, only occurs within xenoliths of argillacious sediments in igneous intrusions

Physical Properties

Name: Muscovite

Commodity: Mica

Formula: KAl2(Si3Al)O10(OH,F)2

Crystal System: Monoclinic ps. hex.

Color: Colorless, grey, green, yellow, brown, violet, rose-red, deep ruby-red, pinkish red

Opacity: Transparent to translucent

Luster: Vitreous to pearly or silky

Streak: Colorless

SGLow: 2.77

SGHigh: 2.88

HardnessLow: 2.5

HardnessHigh: 4

Cleavage: 1

Direction: {001} perfect

Habit: Crystals tabular, hexagonal or diamond-shaped cross section; massive, scaly or lamellar; plumose or stellate aggregates

Fracture: Thin laminae flexible and elastic

Other: Used for insulating in electrical equipment, in lubricants, wall finishes, artificial stone

Comments: Widely distributed in igneous rocks, particularly granites and pegmatites and metamorphic schists and gneisses. Also in sedimentary sandstones and mudstones

Optical Properties

Name: Muscovite

Group: Mica

Formula: KAl2(Si3Al)O10(OH,F)2

Crystal System: Monoclinic ps. hex.

Color: Colorless to pale green

Form: Thin, tabular crystals and scaly aggregates. Microcrystalline variety named sericite

Relief: Slight, n>balsam

Birefringence: Strong, 0.037-0.041

2V: 30"-40"

Nalpha or Nord.: 1.556-1.570

NBeta or Nextr.: 1.587-1.607

NGamma: 1.593-1.611

Optical Sign: Biaxial negative

Orientation: Length slow

Pleochroism: Some varieties show pleochroism

Twinning: According to mica law {110}

Cleavage: One perfect {001}

Extinction: Parallel up to 3"

Alteration:

Features: Cleavage, tabular cystals and strong birefringence distinctive. Similar to talc, but axial angle larger, physical examination should differentiate.

Occurrence: Widespread within felsic igneous rocks, gneisses, schists and low grade metamorphic rocks

Physical Properties
Name: Naumannite
Commodity: Selenium (Se)
Formula: Ag2Se
Crystal System: Orthorhombic ps. cub.
Color: greyish iron-black, tarnishes to iridescent brownish
Opacity: Opaque
Luster: Metallic
Streak: Iron-black
SGLow: 7.69
SGHigh: 7.79
HardnessLow: 2.5
HardnessHigh: 2.5
Cleavage: 0
Direction:
Habit: Crystals cubic; massive granular; thin plates
Fracture: Hackly; sectile and malleable
Other: Used in fade resistant paint pigments, glass production, chemical industry and photo-electric instruments. Alloyed with copper and steel to improve workability
Comments: Selenium minerals are associated with sulphides in metal sulphide ore deposits. Recovered as a byproduct during copper sulphide ore smelting

Optical Properties
Name: Naumannite
Commodity: Selenium (Se)
Formula: Ag2Se
Crystal System: Orthorhombic ps. cub.
Color:
Pleochroism:
Anisotropism:
Internal Reflections:
Reflectance 546nm:
Reflectance at 589nm:
Quant. Color Coordinates (QC):
Vickers Hardness Number (100g):
Polishing Hardness:
Cleavage:
Form: Crystals cubic; massive granular; thin plates
Alteration:
Associated Minerals:
Distinguishing Features:
Model: Selenium minerals are associated with sulphides in metal sulphide ore deposits. Recovered as a byproduct during copper sulphide ore smelting

Physical Properties

Name: Niccolite (Nickeline)
Commodity: Nickel (Ni)
Formula: NiAs
Crystal System: Hexagonal
Color: Copper-red
Opacity: Opaque
Luster: Metallic
Streak:
SGLow: 7.77
SGHigh: 7.78
HardnessLow: 5
HardnessHigh: 5.5
Cleavage:
Direction:
Habit:
Fracture:
Other: Iron alloy for making stainless steel due to corrosion resistence, also non-ferrous alloys. Non toxic, used for food handling and pharmaceutical equipment. Used for electroplating steels and as a base for chromium plate
Comments: Occurs in ultramfic and mafic igneous rocks with chalcopyrite, pyrrhotite and nickel sulphides. Alsi in vein deposits with silver, silver-arsenic and cobalt ores

Optical Properties

Name: Niccolite (Nickeline)
Commodity: Nickel (Ni)
Formula: NiAs
Crystal System: Hexagonal
Color: Yellowish pink to brownish pink
Pleochroism: Strong, yellowish pink to brownish pink
Anisotropism: Very strong, yellow to greenish violet blue to blue Gray
Internal Reflections: None
Reflectance 546nm: 47.2-51.6
Reflectance at 589nm: 53.3-56.0
Quant. Color Coordinates (QC): 7.77
Vickers Hardness Number (100g): 363-372
Polishing Hardness: Greater than chalcopyrite, equal to breithauptite, less than skutterudite, pyrite
Cleavage:
Form: Subhedral to euhedral crystals and anhedral aggregates, concentric bandss and intergrowths. Twinning and radial aggregates common
Alteration: Intergrowths with arsenides
Associated Minerals: With arsenides
Distinguishing Features: Color, pleochrosim and anisotropy distinctive
Model: Occurs in ultramfic and mafic igneous rocks with chalcopyrite, pyrrhotite and nickel sulphides. Alsi in vein deposits with silver, silver-arsenic and cobalt ores

Physical Properties

Name: Niter

Commodity: Potash

Formula: KNO_3

Crystal System: Orthorhombic

Color: Colorless, white

Opacity: Transparent

Luster: Vitreous

Streak:

SGLow: 2

SGHigh: 2.11

HardnessLow:

HardnessHigh: 2

Cleavage: 2.11

Direction: 3

Habit: Conchoidal to uneven; brittle

Fracture: {011} nearly perfect, {010} good, {110} imperfect

Other: Used in fertilisers and potassium salts. Niter used in manufacture of explosives

Comments: Occurs in soils in arid climates with halite, gypsum and nitratine

Physical Properties

Name: Orpiment

Commodity: Arsenic (As)

Formula: As2S3

Crystal System: Monoclinic

Color: Yellow, yellow-orange, brownish yellow

Opacity: Transparent to translucent

Luster: Resinous, pearly on cleavage

Streak: Pale yellow

SGLow: 3.49

SGHigh: 3.49

HardnessLow: 1.5

HardnessHigh: 2

Cleavage: 2

Direction: {010} perfect, {100} traces

Habit: Crystals short prismatic, elongated on c-axis; granular, foliated or powdery aggregates; columnar, reniform

Fracture: Sectile; cleavage lamellae flexible, inelastic

Other: Used in insecticides and as alloy with copper and lead

Comments: Usually regarded as a gangue mineral. Occurs in epithermal to mesothermal quartz vein deposits with tin, tungsten, gold, silver, sphalerite and pyrite

Optical Properties

Name: Orpiment

Commodity: Arsenic (As)

Formula: As2S3

Crystal System: Monoclinic

Color: Gray

Pleochroism: Strong, white to dull Gray to reddish to dull Gray white

Anisotropism: Strong in oil, masked by internal reflections

Internal Reflections: Intense, white to yellow

Reflectance 546nm: 23.0-27.5

Reflectance at 589nm: 22.1-26.7

Quant. Color Coordinates (QC): 0.294, 0.296, 27.6

Vickers Hardness Number (100g): 22-58

Polishing Hardness: Greater than realgar

Cleavage: {010} perfect, {100} traces

Form: Tabular aggregates and needle-lathlike crystals

Alteration:

Associated Minerals: With stibnite, arsenopyrite, arsenic, pyrite, enargite, sphalerite, loellingite

Distinguishing Features: Internal reflections and anisotropy distinctive

Model: Usually regarded as a gangue mineral. Occurs in epithermal to mesothermal quartz vein deposits with tin, tungsten, gold, silver, sphalerite and pyrite

Physical Properties

Name: Orthoclase

Commodity: Feldspar

Formula: KAlSi3O8

Crystal System: Monoclinic

Color: Colorless, white, grey, yellow, reddish, greenish

Opacity: Transparent to translucent

Luster: Vitreous to pearly

Streak: White

SGLow: 2.55

SGHigh: 2.63

HardnessLow: 6

HardnessHigh: 6.5

Cleavage: 6

Direction: {001} and {010} perfect, {100}, {110}, {-110} and {-201} partings

Habit: Crystals short prismatic, blocky, may have orthorhombic or tetragonal aspect; massive, cleavable to granular; twinned

Fracture: Conchoidal to uneven; brittle

Other: Manufacture of porcelain, pottery and glass. Used for glazes in pottery and as mild abrasive

Comments: Important rock forming mineral widely distributed in igneous and metamorphic rocks. Also present in pegmatites

Optical Properties

Name: Orthoclase

Group: Feldspar

Formula: KAlSi3O8

Crystal System: Monoclinic

Color: Colorless

Form: Subhedral and anhedral coarse to fine crystals, phenocrysts and in spherulites

Relief: Low, n<balsam

Birefringence: Weak, 0.008

2V: 69"-72"

Nalpha or Nord.: 1.518

NBeta or Nextr.: 1.524

NGamma: 1.526

Optical Sign: Biaxial negative

Orientation: Legth fast, r>v

Pleochroism:

Twinning: Simple twinning according to carlsbad law

Cleavage: Three directions, perfect {001}, distinct {010}, imperfect {110}

Extinction: Parallel on {001}, oblique 5"-12" on {010}, increasing with soda content

Alteration: Altered to kaolinite. Dimorphic with sanidine

Features: Distinguished from sanidine by large axial angle. Twinning distinctive.

Occurrence: Widespread within felsic igneous rocks, metamorphic gneisses and sedimentary sandstones

Physical Properties

Name: Osmiridium

Commodity: Osmiridium (Os) and Iridium (Ir)Platinum-iridium u

Formula: Alloy of Os-Ir

Crystal System:

Color:

Opacity:

Luster: Metallic

Streak:

SGLow:

SGHigh:

HardnessLow:

HardnessHigh:

Cleavage:

Direction:

Habit:

Fracture:

Other:

Comments: Occurs in some gold and platinum ores in which it sometimes occurs in small amounts

Physical Properties

Name: Patronite

Commodity: Vanadium (V)

Formula: VS4

Crystal System: Monoclinic

Color:

Opacity:

Luster:

Streak:

SGLow:

SGHigh:

HardnessLow:

HardnessHigh:

Cleavage:

Direction:

Habit:

Fracture:

Other: Alloyed with steel to produce high strength steels including tool steels, oil and gas pipelines, structural steels. Used in chemical and oil industries as catalysts and glass colouring agent

Comments: Occurs with nickel and molybdenum sulphides

Physical Properties

Name: Pentlandite
Commodity: Nickel (Ni)
Formula: (Fe,Ni)9S8
Crystal System: Cubic
Color: Light bronze-yellow
Opacity: Opaque
Luster: Metallic
Streak: Bronze brown
SGLow: 4.6
SGHigh: 5.0
HardnessLow: 3.5
HardnessHigh: 4
Cleavage: 1
Direction: {111} parting
Habit: Massive, granular
Fracture: Conchoidal; brittle
Other: Important Iron alloy for making stainless steel due to corrosion resistence, also non-ferrous alloys. Non toxic, used for food handling and pharmaceutical equipment. Used for electroplating steels and as a base for chromium plate
Comments: Associated with pyrrhotite, cobalt, millerite, selenium, silver and platinum ores. Occurs in Archean ultramafic volcanics (komattiites) in Archean Greenstone Belts, and other ultramafic intrusions

Optical Properties

Name: Pentlandite
Commodity: Nickel (Ni)
Formula: (Fe,Ni)9S8
Crystal System: Cubic
Color: Light creamy to yellowish
Pleochroism: None
Anisotropism: Isotropic
Internal Reflections: None
Reflectance 546nm: 46.5
Reflectance at 589nm: 49.0
Quant. Color Coordinates (QC): 0.332, 0.339, 46.9
Vickers Hardness Number (100g): 268-285, slightly fractured
Polishing Hardness: Greater han chalcopyrite, less than pyrrhotite
Cleavage: {111} parting
Form: Cubic crystals and as veinlets, flames or lamellae in pyrrhotite, chalcopyrite, less common.
Alteration: Alters to violarite and millerite along cracks and grain boundaries
Associated Minerals: With pyrrhotite, magnetite, pyrite, cubanite, mackinawite
Distinguishing Features: Cubic crystals, color and occurrence
Model: Associated with pyrrhotite, cobalt, millerite, selenium, silver and platinum ores. Occurs in Archean ultramafic volcanics (komattiites) in Archean Greenstone Belts, and other ultramafic intrusions

Physical Properties

Name: Phlogopite

Commodity: Mica

Formula: KMg3Si3AlO10(F,OH)2

Crystal System: Monoclinic

Color: Yellowish brown to brownish red, colorless, white, greenish

Opacity: Transparent to translucent

Luster: Pearly, submetallic on cleavage

Streak: Colorless

SGLow: 2.76

SGHigh: 2.90

HardnessLow: 2

HardnessHigh: 2.5

Cleavage: 1

Direction: {001} perfect

Habit: Crystals prismatic, tapered; plates; scales

Fracture: Thin laminae flexible and elastic, tough

Other: Used for insulating in electrical equipment, in lubricants, wall finishes, artificial stone

Comments: Occurs in metamorphosed limestones and igneous rocks rich in magnesium

Optical Properties

Name: Phlogopite

Group: Mica

Formula: KMg3Si3AlO10(F,OH)2

Crystal System: Monoclinic

Color: Colorless to pale brown

Form: Short prismatic and thick six sided tabular crystals

Relief: Fair, n>balsam

Birefringence: Strong, 0.044-0.047

2V: 0"-10"

Nalpha or Nord.: 1.551-1.562

NBeta or Nextr.: 1.598-1.606

NGamma: 1.598-1.606

Optical Sign: Biaxial negative

Orientation: Length slow, r>v weak

Pleochroism: Weak pleochroism

Twinning: According to mica law

Cleavage: One perfect {001}

Extinction: Parallel, upto 5"

Alteration:

Features: Similar to biotite but lighter color and weaker adsorption. Similar to muscovite but smaller axial angle. occurrence may be distinctive.

Occurrence: Characteristic of metamorphic limestones

Physical Properties

Name: Phosphate Rock

Commodity: Phosphates

Formula:

Crystal System:

Color:

Opacity:

Luster:

Streak:

SGLow:

SGHigh:

HardnessLow:

HardnessHigh:

Cleavage:

Direction:

Habit:

Fracture:

Other: Fertilisers and in manufacture of phosphorous chemicals

Comments: Associated with marine sediments. Also as guano deposits formed from sea bird droppings on oceanic islands

Physical Properties
Name: Plagioclase
Group: Feldspar
Formula: (Na,Ca)(Si,Al)Si3O8
Crystal System: Triclinic
Color: White to colorless, bluish, grey, reddish, greenish
Opacity: Transparent to subtranslucent
Luster: Vitreous, pearly
Streak: White
SGLow: 2.60
SGHigh: 2.63
HardnessLow: 6
HardnessHigh: 6.5
Cleavage: 2
Direction: {001} perfect, {010} nearly perfect
Habit: Crystals tabular, small; usually massive, lamellar, or granular; twinning common, simple, multiple, and repeated
Fracture: Conchoidal to uneven; brittle
Other: Manufacture of porcelain, pottery and glass. Used for glazes in pottery and as mild abrasive
Comments: Important rock forming mineral widely distributed in igneous and metamorphic rocks

Optical Properties
Name: Plagioclase
Group: Feldspar
Formula: (Na,Ca)(Si,Al)Si3O8
Crystal System: Triclinic
Color: Colorless
Form: Plates or lathe shaped sections, rarely in phenocrysts
Relief: Low, n considerably <balsam, n=balsam
Birefringence: Weak, 0.009-0.011
2V: 77"-82"
Nalpha or Nord.: 1.527-1.533
NBeta or Nextr.: 1.531-1.537
NGamma: 1.538-1.542
Optical Sign: Biaxial positive
Orientation:
Pleochroism:
Twinning: Polysynthetic twinning according to albite law. Twinning according to carlsbad law either alone or combined. Occasional pericline twinning
Cleavage: Four directions, perfect {001}, distinct {010}, imperfect {110} and {11_0}
Extinction: Maximum in albite twins is 12'-19". In cleavage sections parallel to {001} 3"-5" and parallel to {010} 15"-20"
Alteration: Alters to clay
Features: RI equal to and less than balsam is distinctive, albite twinning and extinction angles help.
Occurrence: Most abundant in felsic igneous rocks. Widespread occurrence

Physical Properties

Name: Platinum

Commodity: Platinum (Pt)

Formula: Pt

Crystal System:

Color:

Opacity:

Luster:

Streak:

SGLow:

SGHigh:

HardnessLow:

HardnessHigh:

Cleavage:

Direction:

Habit:

Fracture:

Other: Used in dentistry. High melting point and corrosion resistant . Used as electrical contacts, chemical crucibles

Comments: Occurs in small amounts in igneous rocks, associated with chromite and copper ores and in alluvial deposits

Optical Properties

Name: Platinum

Commodity: Platinum (Pt)

Formula: Pt

Crystal System: Cubic

Color: White

Pleochroism: None

Anisotropism: Isotropic

Internal Reflections: None

Reflectance 546nm: 70.3

Reflectance at 589nm: 71.9

Quant. Color Coordinates (QC): 0.318, 0.324, 70.7

Vickers Hardness Number (100g): 122-129, perfect

Polishing Hardness: Greater than sphalerite, less than pyrrhotite

Cleavage: {001} perfect

Form: Subhedral to euhedral cubic crystals, occasionally zoned or with exsolution laths of iridium and osmium

Alteration:

Associated Minerals: With chromite, pyrrhotite, magnetite, pentlandite, chalcopyrite

Distinguishing Features: Crystals prismatic, tapered; plates; scales

Model: Occurs in small amounts in igneous rocks, associated with chromite and copper ores and in alluvial deposits

Physical Properties

Name: Pollucite

Commodity: Cesium (Ce)

Formula: $(Cs,Na)_2Al_2Si_4O_{12}+H_2O$

Crystal System: Cubic

Color: Colorless, white, grey, pinkish, bluish, violet

Opacity: Transparent

Luster: Vitreous to greasy

Streak:

SGLow: 2.94

SGHigh: 2.94

HardnessLow: 6.5

HardnessHigh: 7

Cleavage: 0

Direction:

Habit: Crystals cubic, dodecahedral, rare; massive, fine-grained

Fracture: Conchoidal to uneven; brittle

Other: Used in photoelectric cells, spectrophotometers, infra red detectors and photomultiplier tubes

Comments: Occurs in pegmatites. Rare.

Optical Properties

Name: Polybasite

Commodity: Silver (Ag)

Formula: Ag16Sb2S11

Crystal System: Trigonal or hexagonal

Color: Gray

Pleochroism: Weak in air, distinct in oil, green to Gray with violet tint

Anisotropism: Moderate in air, strong in oil, blue Gray to yellow green to brown

Internal Reflections: Deep red

Reflectance 546nm: 30.7-32.5

Reflectance at 589nm: 30.0-31.4

Quant. Color Coordinates (QC): 0.302, 0.314, 32.2

Vickers Hardness Number (100g):

Polishing Hardness: Greater than argentite, equal to pyrargyrite, less than stephanite

Cleavage: {10-11} distinct

Form: Crystals prismatic, rhombohedral or scalenohedral, highly modified; massive; compact; disseminated; crusts

Alteration: Solid solution with pearcite

Associated Minerals: With other Sb minerals

Distinguishing Features: Crystals prismatic, rhombohedral or scalenohedral, highly modified; massive; compact; disseminated; crusts. Pleochroism, anisotropy and internal reflections

Physical Properties

Name: Proustite

Commodity: Silver (Ag)

Formula: Ag3AsS3

Crystal System: Trigonal

Color: Scarlet red to vermillion red

Opacity: Transparent to translucent

Luster: Adamantine to submetallic

Streak: Bright red

SGLow: 5.55

SGHigh: 5.64

HardnessLow: 2

HardnessHigh: 2.5

Cleavage: 1

Direction: {10-11} distinct

Habit: Crystals prismatic, rhombohedral or scalenohedral, highly modified; massive; compact; disseminated; crusts

Fracture: Conchoidal to uneven; brittle

Other: Used in coinage, jewellery, sterling ware, mirrors, electroplating, batteries and photographic and electronic products

Comments:

Optical Properties

Name: Proustite

Commodity: Silver (Ag)

Formula: Ag3AsS3

Crystal System: Trigonal or hexagonal

Color: Bluish Gray

Pleochroism: Distinct, yellowish to bluish Gray

Anisotropism: Strong, masked by internal reflections

Internal Reflections: Scarlet red

Reflectance 546nm: 27.4-28.1

Reflectance at 589nm: 25.8-26.4

Quant. Color Coordinates (QC): 0.287, 0.291, 28.1

Vickers Hardness Number (100g): 103-137 50g

Polishing Hardness: Equal to pyrargyrite

Cleavage:

Form: Crystals prismatic

Alteration: Solid solution with pyrargyrite

Associated Minerals: With other As minerals, similar to pyrargyrite

Distinguishing Features: Pleochroism, anisotropy and internal reflections

Physical Properties

Name: Psilomelane

Commodity: Manganese (Mn)

Formula:

Crystal System:

Color:

Opacity:

Luster:

Streak:

SGLow:

SGHigh:

HardnessLow:

HardnessHigh:

Cleavage:

Direction:

Habit:

Fracture:

Other: Important iron alloy for steel making

Comments: Occurs in veins with other manganese minerals and with pyrolusite and limonite in sediments

Optical Properties

Name: Psilomelane

Commodity: Manganese (Mn)

Formula: MnO

Crystal System: Monoclinic

Color: Bluish Gray to Grayish white

Pleochroism: Strong, white to bluish Gray

Anisotropism: Strong, white to Gray

Internal Reflections: Occassional brown

Reflectance 546nm: 15-30

Reflectance at 589nm:

Quant. Color Coordinates (QC):

Vickers Hardness Number (100g): 203-813

Polishing Hardness:

Cleavage:

Form: Botryoidal and acicular concentric aggregates. Often intergrown with pyrolusite and cryptomelane

Alteration:

Associated Minerals: With other Mn oxides

Distinguishing Features: Pleochroism, anisotropy and occurrence

Model: Occurs in veins with other manganese minerals and with pyrolusite and limonite in sediments

Physical Properties

Name: Pyrargyrite

Commodity: Silver (Ag)

Formula: Ag3SbS3

Crystal System: Trigonal

Color: Deep red

Opacity: Translucent

Luster: Adamantine

Streak: Dark red

SGLow: 5.85

SGHigh: 5.85

HardnessLow: 2.5

HardnessHigh: 2.5

Cleavage: 2

Direction: {10-11} distinct, {01-12} indistinct

Habit: Crystals prismatic, hemimorphism sometimes distinct, or steep scalenohedral; massive, compact; disseminated or crusts

Fracture:

Other: Used in coinage, jewellery, sterling ware, mirrors, electroplating, batteries and photographic and electronic products

Comments: Associated with lead, zinc and copper ores. Recovered as a byproduct during smelting of these ores

Optical Properties

Name: Pyrargyrite

Commodity: Silver (Ag)

Formula: Ag3SbS3

Crystal System: Trigonal or hexagonal

Color: Bluish Gray

Pleochroism: Distinct to strong, bluish Gray

Anisotropism: Strong, in oil, Gray to dark Gray, masked by internal reflections

Internal Reflections: Intense red

Reflectance 546nm: 28.5-30.3

Reflectance at 589nm: 26.5-28.4

Quant. Color Coordinates (QC): 0.287, 0.244, 30.2

Vickers Hardness Number (100g): 66-87, parallel to cleavage

Polishing Hardness: Greater than polybasite, less than galena

Cleavage: {10-11} distinct, {01-12} indistinct

Form: Irregular grains and aggregates, sometimes twinned and zoned

Alteration:

Associated Minerals: With galena, Sb sulfosalts, pyrite, sphalerite, chalcopyrite, tetrahedrite, arsenopyrite, Ni-Co-Fe arsenides

Distinguishing Features: Pleochroism, internal reflections

Model: Associated with lead, zinc and copper ores. Recovered as a byproduct during smelting of these ores

Physical Properties
Name: Pyrite
Commodity: Gold (Au)
Formula: FeS2
Crystal System: Cubic
Color: Brass-yellow
Opacity: Opaque
Luster: Metallic
Streak:
SGLow: 5.018
SGHigh: 5.028
HardnessLow: 6
HardnessHigh: 6.5
Cleavage: 3
Direction: {100} and {311} indistinct, {110} occasional parting
Habit: Crystals cubic, pyritohedral, octahedral and combinations, abnormally developed, rarely acicular; massive; granular
Fracture: Conchoidal to uneven; brittle
Other: Monetary use, jewellery, electronics and for decoration
Comments: Associated with other sulphide minerals and gold in hypothermal to epithermal quartz veins. Sometimes contains small amounts of gold and is mined as an ore of gold

Optical Properties
Name: Pyrite
Commodity: Gold (Au)
Formula: FeS2
Crystal System: Cubic
Color: Yellowish white to brass yellow
Pleochroism: None
Anisotropism: Weak, blue green to orange red
Internal Reflections: None
Reflectance 546nm: 51.7
Reflectance at 589nm: 53.5
Quant. Color Coordinates (QC): 0.327, 0.335, 51.8
Vickers Hardness Number (100g): 150-162, fractured
Polishing Hardness: Greater than arsenopyrite, marcasite, less than cassiterite
Cleavage: {100} and {311} indistinct, {110} occasional parting
Form: Euhedral cubes and pyritohedra, anhedral crystalline masses and colloform bands. Zoning and twinning sometimes
Alteration:
Associated Minerals: With most ore types
Distinguishing Features: Crystals cubic, pyritohedral, octahedral and combinations, abnormally developed, rarely acicular; massive; granular. Color and abundance distinctive
Model: Associated with other sulphide minerals and gold in hypothermal to epithermal quartz veins. Sometimes contains small amounts of gold and is mined as an ore of gold

Physical Properties

Name: Pyrochlore

Commodity: Tantalum (Ta)

Formula: (Ca,Na)2Nb2O6(OH,F)

Crystal System: Cubic

Color: Yellowish brown, reddish brown, brown to black

Opacity: Translucent to opaque

Luster: Vitreous to resinous

Streak:

SGLow: 4.48

SGHigh: 4.48

HardnessLow: 5

HardnessHigh: 5.5

Cleavage: 1

Direction: {111} sometimes distinct, or parting

Habit: Crystals octahedral, to 1 cm; irregular masses; embedded grains

Fracture: Subconchoidal to uneven; brittle

Other: Very high corrosion resistance. Used in production of special steels for medical use, in chemical and electrical processes, electrodes, cutting tools and manufacture of electrodes

Comments: Associated with zircon and apatite in pegmatites. Also recovered from tin slags as a byproduct

Physical Properties

Name: Pyrolusite

Commodity: Manganese (Mn)

Formula: MnO_2

Crystal System: Tetragonal

Color: Black, dark steel-grey

Opacity: Opaque

Luster: Metallic, dull

Streak: Black, bluish black

SGLow: 5.06

SGHigh: 5.06

HardnessLow: 6

HardnessHigh: 6.5

Cleavage: 1

Direction: {110} perfect

Habit: Crystals short to long prismatic, elongated along c-axis, or equant; massive, compact, columnar, fibrous; stalactitic

Fracture: Uneven; brittle

Other: Important iron alloy for steel making

Comments: In oxidized zone of ore deposits and quartz veins. Also deposited as nodules on the ocean floor

Optical Properties

Name: Pyrolusite

Commodity: Manganese (Mn)

Formula: MnO_2

Crystal System: Tetragonal

Color: Creamy white

Pleochroism: Distinct in oil, yellowish white to Gray white

Anisotropism: Very strong, yellowish to brownish to blue

Internal Reflections: None

Reflectance 546nm: 29.0-40.0

Reflectance at 589nm: 28.1-39.3

Quant. Color Coordinates (QC):

Vickers Hardness Number (100g): 146-243, fractured

Polishing Hardness: Variable

Cleavage: {110} perfect

Form: Coarse grained tabular crystals and banded aggregates, occasionally twinned

Alteration:

Associated Minerals: With other Mn oxides

Distinguishing Features: Color, pleochroism, anisotropy

Model: In oxidized zone of ore deposits and quartz veins. Also deposited as nodules on the ocean floor

Physical Properties

Name: Pyrope

Commodity: Garnet

Formula: Mg3Al2(SiO4)3

Crystal System: Cubic

Color: Pinkish red, purplish red, orange-red, crimson, nearly black

Opacity: Transparent to translucent

Luster: Vitreous

Streak:

SGLow: 3.5

SGHigh: 3.8

HardnessLow: 7

HardnessHigh: 7.5

Cleavage: 1

Direction: {110} parting sometimes distinct

Habit: Crystals dodecahedrons or trapezohedrons, rare; rounded pebbles; embedded grains

Fracture: Conchoidal; brittle

Other: Abrasives, gemstones

Comments: Occurs in many types of metamorphic rocks and some igneous rocks and in alluvial and beach deposits

Optical Properties

Name: Pyrope

Group: Garnet

Formula: Mg3Al2(SiO4)3

Crystal System: Isometric, cubic

Color: Colorless

Form: Euhedral dodecahedrons in six sided trapezohedrons in eigth sided cross sections. Plygonal grains and aggregates.

Relief: V high, n>balsam

Birefringence: Isotropic, may show weak birefringence

2V:

Nalpha or Nord.: 1.741-1.760

NBeta or Nextr.:

NGamma:

Optical Sign:

Orientation:

Pleochroism:

Twinning:

Cleavage: May have parting parallel to {110}, irregular fractures

Extinction:

Alteration:

Features: Similar to spinel, which is octohedral.

Occurrence: Mainly occurs within peridotites and metamorphic serpentinites

Physical Properties

Name: Pyrrhotite

Commodity: Iron (Fe)

Formula: Fe1-xS

Crystal System: Monoclinic and hexagonal

Color: Bronze-yellow to bronze-red; tarnishes dark brown

Opacity: Opaque

Luster: Metallic

Streak: greyish black

SGLow: 4.53

SGHigh: 4.77

HardnessLow: 3.5

HardnessHigh: 4.5

Cleavage: 1

Direction: Basal parting may be distinct

Habit: Crystals platy to tabular or bipyramidal, well-formed; massive, granular aggregates

Fracture: Subconchoidal to uneven; brittle

Other: Iron and Steel

Comments: Associated with other sulphide minerals and gold in hypothermal to epithermal quartz veins and with nickel minerals in nickel sulphide deposits in ultramafic rocks

Optical Properties

Name: Pyrrhotite

Commodity: Iron (Fe)

Formula: Fe1-xS

Crystal System: Monoclinic and hexagonal

Color: Creamy pinkish brown

Pleochroism: Distinct, creamy brown to reddish brown

Anisotropism: Very strong, yellow Gray to Grayish blue

Internal Reflections: None

Reflectance 546nm: Hex 34.0-39.2 Mono 34.8-39.9

Reflectance at 589nm: Hex 35.8-40.7 Mono 36.9-41.6

Quant. Color Coordinates (QC): Mono 0.330, 0.334, 35.3

Vickers Hardness Number (100g): 230-259, perfect (anisotropic sections), 280-318, perfect isotropic sections, 373-409, perfect, monoclinic variety

Polishing Hardness: Greater than chalcopyrite, equal to pentlandite, less than pyrrhotite

Cleavage: Basal parting may be distinct

Form: Anhedral granular aggregates frequent. May be platy to tabular crystals. Lamellar exsolution of mono and hex forms common

Alteration: Alters to a rim of mono pyrrhotite during weathering.

Associated Minerals: With most sulphide minerals.

Distinguishing Features: Creamy brown color and pleochroism distinctive

Model: Associated with other sulphide minerals and gold in hypothermal to epithermal quartz veins and with nickel minerals in nickel sulphide deposits in ultramafic rocks

Physical Properties

Name: Quartz (alpha-Quartz)

Commodity: Silicon (Si)

Formula: SiO2

Crystal System: Trigonal

Color: Colorless, white, grey, yellow to brown to black, violet, pink

Opacity: Transparent to translucent

Luster: Vitreous, sometimes greasy or waxy

Streak: White

SGLow: 2.655

SGHigh: 2.655

HardnessLow: 7

HardnessHigh: 7

Cleavage: 0

Direction: ({10-11}, {01-11}, {10-10}, {0001},{11-20}, {11-21} and {51-61} seldom distinct)

Habit: Crystals short to long prismatic, elongated along c-axis, hexagonal, horizontally striated, bent, distorted, skeletal

Fracture: Subconchoidal to conchoidal; brittle

Other: Alloyed with iron to make heavy-medium ferro-silicon. Also used as a semi-conductor

Comments: Most common mineral in the earth's crust. Occurs in beach sand and quartzite deposits

Optical Properties

Name: Quartz

Group: Quartz

Formula: SiO2

Crystal System: Hexagonal

Color: Colorless

Form: Euhedral hexagonal crystals, anhedral grains, veins, deisseminated grains. Anhedral fne to coarse crystals

Relief: V.low, n>balsam

Birefringence: Weak, 0.009

2V:

Nalpha or Nord.: 1.5442

NBeta or Nextr.: 1.5533

NGamma:

Optical Sign: Uniaxial positive

Orientation: Length slow in euhedral crystals

Pleochroism:

Twinning: Common, but rarely shows in thin section

Cleavage: None

Extinction: Parallel in euhedral crystals, wavy extinction in large crsytals due to strain.

Alteration:

Features: Absence of cleavage, hexagonal crystals, first order birefringence, uniaxial positive

Occurrence: Most widespread of all minerals

Physical Properties
Name: Rammelsbergite
Commodity: Nickel (Ni)
Formula: NiAs2
Crystal System: Orthorhombic
Color: Tin-white with pinkish tint
Opacity: Opaque
Luster: Metallic
Streak:
SGLow: 6.97
SGHigh: 6.97
HardnessLow: 5.5
HardnessHigh: 6
Cleavage:
Direction:
Habit: Crystals short prismatic, to 1 mm; massive, fine granular to prismatic or radial-fibrous structure.
Fracture: Uneven
Other: Important Iron alloy for making stainless steel due to corrosion resistence, also non-ferrous alloys. Non toxic, used for food handling and pharmaceutical equipment. Used for electroplating steels and as a base for chromium plate
Comments: Associated with pyrrhotite, cobalt, millerite, selenium, silver and platinum ores. Occurs in Archean ultramafic volcanics (komattiites) in Archean Greenstone Belts, and other ultramafic intrusions

Optical Properties
Name: Rammelsbergite
Commodity: Nickel (Ni)
Formula: NiAs2
Crystal System: Orthorhombic
Color: White
Pleochroism: Weak in air, distinct in oil, yellowish to bluish
Anisotropism: Strong, pinkish to brownish to greenish to bluish
Internal Reflections: None
Reflectance 546nm: 53.2-56.3
Reflectance at 589nm: 53.5-56.1
Quant. Color Coordinates (QC):
Vickers Hardness Number (100g): 585-803, slightly fractured
Polishing Hardness: Equal to skutterudite, less than safflorite, loellingite.
Cleavage:
Form: Fine grained aggregates of interlocking crystals, often zoned, spherulitic, radiating and fibrous. Simple or complex twinning
Alteration: Intergrowths with niccolite and Co-Ni-Fe arsenides
Associated Minerals: With other Ni-Co arsenides and niccolite
Distinguishing Features: White color, pleochroism and anisotropy distinctive
Model: Associated with pyrrhotite, cobalt, millerite, selenium, silver and platinum ores. Occurs in Archean ultramafic volcanics (komattiites) in Archean Greenstone Belts, and other ultramafic intrusions

Physical Properties

Name: Realgar

Commodity: Arsenic (As)

Formula: AsS

Crystal System: Monoclinic

Color: Orange-red to dark red

Opacity: Transparent to translucent

Luster: Resinous to greasy

Streak: Orange-yellow

SGLow: 3.56

SGHigh: 3.56

HardnessLow: 1.5

HardnessHigh: 2

Cleavage: 4

Direction: {010} distinct; {-101}, {100} and {120} distinct

Habit: Crystals short prismatic, striated parallel to c-axis; massive, compact, coarse- to fine-grained; powdery coatings

Fracture: Conchoidal; sectile

Other: Used in insecticides and as alloy with copper and lead

Comments: Usually regarded as a gangue mineral. Occurs in epithermal to mesothermal quartz vein deposits with tin, tungsten, gold, silver, sphalerite and pyrite

Optical Properties

Name: Realgar

Commodity: Arsenic (As)

Formula: AsS

Crystal System: Monoclinic

Color: Dull Gray

Pleochroism: Weak, Gray with reddish to bluish tint

Anisotropism: Strong, masked by internal reflections

Internal Reflections: Strong, yellowish red

Reflectance 546nm: 22.1

Reflectance at 589nm: 20.9

Quant. Color Coordinates (QC):

Vickers Hardness Number (100g): 47-60

Polishing Hardness: Less than orpiment

Cleavage: {010} distinct; {-101}, {100} and {120} distinct

Form: Irregular platy aggregates with orpiment

Alteration:

Associated Minerals: With other As sulphides, stibnite, arsenopyrite, pyrite, arsenic, tennantite, enargite, proustite

Distinguishing Features: Yellowish red internal reflections characteristic

Model: Usually regarded as a gangue mineral. Occurs in epithermal to mesothermal quartz vein deposits with tin, tungsten, gold, silver, sphalerite and pyrite

Physical Properties

Name: Roscoelite

Commodity: Vanadium (V)

Formula: K(V,Al,Mg)2AlSi3O10(OH)2

Crystal System: Monoclinic

Color: Clove-brown to greenish brown, dark green

Opacity: Translucent

Luster: Pearly

Streak:

SGLow: 2.97

SGHigh: 2.97

HardnessLow: 2.5

HardnessHigh: 2.5

Cleavage: 1

Direction: {001} perfect

Habit: Minute scales, in stellate groups

Fracture:

Other: Alloyed with steel to produce high strength steels including tool steels, oil and gas pipelines, structural steels. Used in chemical and oil industries as catalysts and glass colouring agent

Comments: Secondary mineral in sedimentary rocks and also associated with uraninite in uraninite deposits

Physical Properties

Name: Rutile

Commodity: Titanium (Ti)

Formula: TiO2

Crystal System: Tetragonal

Color: Reddish brown to red, yellow, orange-yellow, bluish, greyish black to black, greenish

Opacity: Transparent to translucent

Luster: Adamantine, submetallic

Streak: Pale brown to yellowish

SGLow: 4.23

SGHigh: 4.23

HardnessLow: 6

HardnessHigh: 6.5

Cleavage: 5

Direction: {110} and {100} distinct, {111} traces, {092} and {011} parting

Habit: Crystals short prismatic, often striated along c-axis; slender prismatic to acicular; massive, compact to granular

Fracture: Conchoidal to uneven; brittle

Other: High strength and corrosion resistance. Mostly used in aircraft and aerospace industries. Used in chemical and desalination plants and heatbexchanger tubing in power plants

Comments: Associated with monazite, zircon, ilmenite etc in alluvial sands and heavy mineral strandline beach deposits. An accessory mineral in igneous rocks concentrated by sedimentary processes

Optical Properties

Name: Rutile

Commodity: Titanium (Ti)

Formula: TiO2

Crystal System: Tetragonal

Color: Gray, faint bluish tint

Pleochroism: Distinct pale brown to yellowish

Anisotropism: Strong, masked by internal reflections

Internal Reflections: Strong, white, yellowish, redish brown

Reflectance 546nm: 20.3

Reflectance at 589nm: 19.8

Quant. Color Coordinates (QC):

Vickers Hardness Number (100g): 1132-1187, perfect

Polishing Hardness: Greater than ilmenite, less than hematite

Cleavage: {110} and {100} distinct, {111} traces, {092} and {011} parting

Form: Subhedral to euhedral columnar and needle like crystals

Alteration:

Associated Minerals: With hematite, ilmenite, tantalite

Distinguishing Features: Internal reflections distinctive

Model: Associated with monazite, zircon, ilmenite etc in alluvial sands and heavy mineral strandline beach deposits. An accessory mineral in igneous rocks concentrated by sedimentary processes

Physical Properties

Name: Safflorite

Commodity: Cobalt (Co)

Formula: CoAs2

Crystal System: Orthorhombic

Color: Tin-white, tarnishes dark grey

Opacity:

Luster: Metallic

Streak: Greyish black

SGLow: 7.2

SGHigh: 7.2

HardnessLow: 4.5

HardnessHigh: 5

Cleavage: 1

Direction: {100} distinct

Habit: Crystals short to long prismatic, resemble arsenopyrite; massive, radial-fibrous

Fracture: Conchoidal to uneven; brittle

Other: Alloying to produce high temperature steels and magnetic alloys. Used as a catalyst in the chemical industries and as cement in sintered carbide cutting equipment

Comments: Occurs in veins with arsenopyrite, silver, calcite and nickel minerals. Cobalt is extracted as a byproduc during lead, copper and nickel smelting

Optical Properties

Name: Safflorite

Commodity: Cobalt (Co)

Formula: CoAs2

Crystal System: Orthorhombic

Color: White with a bluish tint

Pleochroism: Weak, bluish to Gray

Anisotropism: Strong

Internal Reflections: None

Reflectance 546nm: 55-60

Reflectance at 589nm:

Quant. Color Coordinates (QC):

Vickers Hardness Number (100g): 285-464

Polishing Hardness: Greater than skutterudite, less than loellingite

Cleavage: {100} distinct

Form: Radiating aggregates of anhedral to subhedral crystals and in concentric layers with other arsenide minerals. Also starlike triplets. Often twinned

Alteration:

Associated Minerals: With other arsenide minerals

Distinguishing Features: Crystals short to long prismatic, resemble arsenopyrite; massive, radial-fibrous. Bluish Gray pleochroism and anisotropy

Model: Occurs in veins with arsenopyrite, silver, calcite and nickel minerals. Cobalt is extracted as a byproduc during lead, copper and nickel smelting

Physical Properties

Name: Sanidine

Commodity: Feldspar

Formula: (K,Na)(Si,Al)4O8

Crystal System: Monoclinic

Color: Colorless, whitish, tan, pinkish

Opacity:

Luster: Vitreous

Streak: White

SGLow: 2.56

SGHigh: 2.62

HardnessLow: 6

HardnessHigh: 6

Cleavage: 6

Direction: {001} and {010} perfect; {100}, {110}, {-110} and {-210} partings

Habit: Crystals prismatic, often tabular

Fracture:

Other: Manufacture of porcelain, pottery and glass. Used for glazes in pottery and as mild abrasive

Comments: Important rock forming mineral widely distributed in igneous and metamorphic rocks

Optical Properties

Name: Sanidine

Group: Feldspar

Formula: (K,Na)(Si,Al)4O8

Crystal System: Monoclinic

Color: Colorless

Form: Subhedral to eudral monoclinic prisms as phenocrystals

Relief: Low, n<balsam

Birefringence: Weak, 0.007

2V: 0"-12"

Nalpha or Nord.: 1.517-1.520

NBeta or Nextr.: 1.523-1.525

NGamma: 1.524-1.526

Optical Sign: Biaxial negative

Orientation: (1)axial plane {010} r>v, (2) axial plane perpendicular to {010} r<v

Pleochroism:

Twinning: Simple twinning according to carlsbad law

Cleavage: Two directions, perfect {001}, distinct {010}, may be parting parallel to {100}

Extinction: Parallel on {001}, >5" on {010}, Sections normal to optic axis are almost dark, due to small axial angle

Alteration:

Features: Small axial angle differentaites from orthoclase

Occurrence: Considered characteristic of felsic volcanic rocks. Widespread occurrence

Physical Properties

Name: Sassolite (Sassoline)

Commodity: Borates

Formula: H3BO3

Crystal System: Triclinic

Color: White to grey, discolored yellowish or brownish

Opacity: Transparent

Luster: Pearly

Streak:

SGLow: 1.46

SGHigh: 1.52

HardnessLow: 1

HardnessHigh: 1

Cleavage: 1

Direction: {001} perfect, micaceous

Habit: Crystals tabular, pseudohexagonal; plates <1 mm; acicular, rare; coatings and small scales

Fracture: Crystals flexible

Other: Used in manufacture of insulating fibreglass, as fluxes for manufacture of glass and enamels. Borax is used in cloth manufacture and tanning, soap and glue, as preservatives and antiseptics

Comments: Associated with sulphur in volcanoes and hot springs

Physical Properties

Name: Scheelite

Commodity: Tungsten (W)

Formula: CaWO4

Crystal System: Tetragonal

Color: Colorless, white, grey, brownish, orange-yellow, greenish, purplish

Opacity: Transparent to translucent

Luster: Vitreous to adamantine

Streak:

SGLow: 6.10

SGHigh: 6.10

HardnessLow: 4.5

HardnessHigh: 5

Cleavage: 3

Direction: {101} distinct; {112} interrupted; {001} indistinct

Habit: Crystals octahedral or tabular, often diagonally striated; massive, granular; columnar

Fracture: Subconchoidal to uneven

Other: High wear resistance. Tungsten carbide for cutting and drilling tools. Used in lamp filaments, electronic parts, electrical contacts. Important steel alloy to produce tool steels

Comments: Associated with wolfram, molybdenite and cassiterite in granites and pegmatites. Occurs in contact metamorphic aureoles around granite plutons.

Optical Properties

Name: Scheelite

Commodity: Tungsten (W)

Formula: CaWO4

Crystal System: Tetragonal

Color: Gray white

Pleochroism: None

Anisotropism: Distinct, masked by internal reflections

Internal Reflections: Frequent, white

Reflectance 546nm: 9.9-10

Reflectance at 589nm: 9.7-9.9

Quant. Color Coordinates (QC):

Vickers Hardness Number (100g): 387-409, fractured

Polishing Hardness: Less than wolframite

Cleavage: {101} distinct; {112} interrupted; {001} indistinct

Form: Lathlike polycrystalline aggregates, Partial replacement of wolframite, intergrown with Fe-oxides

Alteration: Partial replacement of wolframite

Associated Minerals: With wolframite, cassiterite

Distinguishing Features: Fluoresces pale blue to yellow under uv. Internal reflections distinctive.

Model: Associated with wolfram, molybdenite and cassiterite in granites and pegmatites. Occurs in contact metamorphic aureoles around granite plutons.

Descriptive Handbook of Economic Minerals

Physical Properties

Name: Serpentine (Serpentinite) (Group Name)
Commodity: Serpentine
Formula: Mg3Si2O5(OH)4
Crystal System: Monoclinic
Color: Yellow, white, grey, green
Opacity: Opaque
Luster: Non metallic
Streak:
SGLow: 2.55
SGHigh: 2.56
HardnessLow: 2.5
HardnessHigh: 2.5
Cleavage: 1
Direction: {001} perfect
Habit: Fibrous, tabular
Fracture: Plates
Other: Used as building and ornamental stone. Fibrous varieties used as asbestos
Comments: Occurs in metamorphic belts as schists and alteration product of ultramafic rocks (alteration of olivine and orthopyroxene). Widely distributed in Archean greenstone belts

Optical Properties

Name: Serpentine
Group: Serpentine
Formula: Mg3Si2O5(OH)4
Crystal System: Monoclinic
Color:
Form:
Relief:
Birefringence:
2V:
Nalpha or Nord.:
NBeta or Nextr.:
NGamma:
Optical Sign:
Orientation:
Pleochroism:
Twinning:
Cleavage:
Extinction:
Alteration:
Features: A group name
Occurrence: A metamorphic mineral principally found in serpentinites

Physical Properties

Name: Siderite

Commodity: Iron (Fe)

Formula: FeCO3

Crystal System: Trigonal

Color: Pale yellowish, pale green, yellowish brown, brown, reddish brown, white

Opacity: Translucent to subtranslucent

Luster: Vitreous, pearly or silky

Streak: White

SGLow: 3.96

SGHigh: 3.96

HardnessLow: 4

HardnessHigh: 4

Cleavage: 1

Direction: {10-11} perfect

Habit: Crystals rhombohedral or tabular, prismatic, scalenohedral; massive, coarse to fine granular; botryoidal, globular

Fracture: Conchoidal to uneven; brittle

Other: Iron and Steel

Comments: Occurs in sedimentary basins where it forms massive deposits in sedimentary rocks

Optical Properties

Name: Siderite

Group: Calcite

Formula: FeCO3

Crystal System: Hexagonal

Color: Colorless to grey, brownish

Form: Anhedral to euhedral fine to coarse aggregates. Oolitic, spherulitic or colloform

Relief: High along long driection, low along short direction

Birefringence: Extreme, 0.234-0.242

2V: 3.96

Nalpha or Nord.: 1.596-1.633

NBeta or Nextr.: 1.830-1.875

NGamma:

Optical Sign: Uniaxial negative

Orientation:

Pleochroism:

Twinning: Polysynthetic parallel to long direction {011_2} occasionally

Cleavage: One perfect {101_1}

Extinction: Symmetrical to cleavage

Alteration:

Features: Similar calcite, dolomite and magnesite. Brown iron staining distinctive. Index of refraction < balsam, others index of refraction > balsam

Occurrence: Occurs as diagenetic cement in sandstones, in veins and replacement deposits and sedimentary oolitic ironstones

Optical Properties

Name: Siegenite

Commodity: Cobalt (Co)

Formula: (Co,Ni)3S4

Crystal System: Trigonal or hexagonal

Color: Creamy white with slight pink tinge

Pleochroism: None

Anisotropism: Isotropic

Internal Reflections: None

Reflectance 546nm: 45.4

Reflectance at 589nm: 47.2

Quant. Color Coordinates (QC): 0.321, 0.323, 46.0

Vickers Hardness Number (100g): 503-525, slightly fractured

Polishing Hardness: Equal to linnaeite

Cleavage: {10-11} perfect

Form: Subhedral to euhedral prismatic and anhedral polycrystalline crystal aggregates

Alteration:

Associated Minerals: With Cu and Cu Fe sulphides, pyrite, uraninite

Distinguishing Features: Prismatic crystals, color and isotropism

Model: With Cu and Cu-Fe sulphide deposits

Physical Properties

Name: Sillimanite

Commodity: Aluminum Silicate

Formula: Al2SiO5

Crystal System: Orthorhombic

Color: Colorless, white, grey, yellowish, brownish, greenish, bluish

Opacity: Transparent to translucent

Luster: Vitreous to silky

Streak: Colorless

SGLow: 3.23

SGHigh: 3.27

HardnessLow: 6.5

HardnessHigh: 7.5

Cleavage: 1

Direction: {010} perfect

Habit: Crystals long prismatic, vertically striated, nearly square cross section; massive, fibrous to somewhat columnar

Fracture: Uneven

Other: High alumina refractory for iron and steel industries and other metal smelters. Used in glass industry and as insulating porcelain

Comments: Widely distributed in schists, gneisses, pegmatites and quartz veins in metamorphic belts. Mostly in high grade schists and gneisses

Optical Properties

Name: Sillimanite

Group: Sillimanite

Formula: Al2SiO5

Crystal System: Orthorhombic

Color: Colorless

Form: Small, often minute prismatic crystals and fibrous felted masses. Nearly square in cross section

Relief: High, n>balsam

Birefringence: Moderate, 0.020-0.023

2V: 20"-30"

Nalpha or Nord.: 1.657-1.661

NBeta or Nextr.: 1.658-1.670

NGamma: 1.6777-1.684

Optical Sign: Biaxial positive

Orientation: Length slow, r>v strong

Pleochroism:

Twinning:

Cleavage: One perfect parallel to {010} but not always apparent

Extinction: Parallel in longitudinal sections and symmetrical in cross sections

Alteration:

Features: Similar to andalusite but is length slow, lacks two cleavage at right angles, stronger birefringence.

Occurrence: Widespread within gneisses, schists, slates, hornfelses and other metamorphic rocks

Physical Properties
Name: Silver
Commodity: Silver (Ag)
Formula: Ag
Crystal System: Cubic
Color: Silver-white, tarnishes grey to black
Opacity: Opaque
Luster: Metallic
Streak:
SGLow: 10.50
SGHigh: 10.50
HardnessLow: 2.5
HardnessHigh: 3
Cleavage: 0
Direction:
Habit: Crystals cubic, octahedral, dodecahedral, in parallel groups, elongated, arborescent, reticulated and as wires; massive; scales
Fracture: Hackly; malleable and ductile
Other: Used in coinage, jewellery, sterling ware, mirrors, electroplating, batteries and photographic and electronic products
Comments: Usually alloyed with copper, gold etc in silver sulphide orebodies. Occurs in veins with chlorargyrite and cerrusite

Optical Properties
Name: Silver
Commodity: Silver (Ag)
Formula: Ag
Crystal System: Cubic
Color: Bright white with creamy tint, tarnishes rapidly
Pleochroism: None
Anisotropism: Isotropic
Internal Reflections: None
Reflectance 546nm: 2.5
Reflectance at 589nm: 3
Quant. Color Coordinates (QC): 0.314, 0.321, 94.2
Vickers Hardness Number (100g): 792-907, slightly fractured
Polishing Hardness: Greater than chalcopyrite, equal to tetrahedrite, less than sphalerite
Cleavage:
Form: Usually in irregular aggregates and veinlets, inclusions and dendrites within arsenides
Alteration: Lamellar intergrowths with allargentum
Associated Minerals: Ag, sulfosalts, arsenides, copper sulphides, bismuth, galena, argentite
Distinguishing Features: Color, isotropism and occurrence
Model: Usually alloyed with copper, gold etc in silver sulphide orebodies. Occurs in veins with chlorargyrite and cerrusite

Physical Properties

Name: Skutterudite

Commodity: Cobalt (Co)

Formula: CoAs3

Crystal System: Cubic

Color: Tin-white, may tarnish iridescent or grey

Opacity: Opaque

Luster: Metallic

Streak: Black

SGLow: 6.1

SGHigh: 6.9

HardnessLow: 5.5

HardnessHigh: 6

Cleavage: 3

Direction: {001} and {111} distinct; {011} trace

Habit: Crystals cubic, octahedral, cubo-octahedral, rarely pyritohedral; reticulated shapes; distorted aggregates; massive

Fracture:

Other: Alloying to produce high temperature steels and magnetic alloys. Used as a catalyst in the chemical industries and as cement in sintered carbide cutting equipment

Comments: Occurs in veins with arsenopyrite, silver, calcite and nickel minerals. Cobalt is extracted as a byproduc during lead, copper and nickel smelting

Optical Properties

Name: Skutterudite

Commodity: Cobalt (Co)

Formula: CoAs3

Crystal System: Cubic

Color: Cream white to Grayish white, often zoned

Pleochroism: None

Anisotropism: Isotropic

Internal Reflections: None

Reflectance 546nm: 54.4

Reflectance at 589nm: 53.8

Quant. Color Coordinates (QC): 0.305, 0.312, 54.3

Vickers Hardness Number (100g): 792-907, slightly fractured

Polishing Hardness: Equal to safflorite, greater than linnaeite, less than arsenopyrite, pyrite

Cleavage: {001} and {111} distinct; {011} trace

Form: Usually radial aggregates of bladed crystals , often zoned. Euhedral cubic crystals.

Alteration:

Associated Minerals: Intergrowths with niccolite, bismuth, other arsenides and Ag-Bi-U mineralization

Distinguishing Features: Prismatic crystals, isotropism and color

Model: Occurs in veins with arsenopyrite, silver, calcite and nickel minerals. Cobalt is extracted as a byproduc during lead, copper and nickel smelting

Physical Properties

Name: Smithsonite
Commodity: Zinc (Zn)
Formula: ZnCO3
Crystal System: Trigonal
Color: White, greyish, yellow, yellowish brown to brown, green, bluish green, blue, pink
Opacity: Transparent to translucent
Luster: Vitreous to pearly
Streak: White
SGLow: 4.30
SGHigh: 4.45
HardnessLow: 4
HardnessHigh: 4.5
Cleavage: 1
Direction: {10-11} nearly perfect
Habit: Crystals rhombohedral, often curved and rough; crusts; botryoidal, stalactitic, reniform; massive, granular to compact
Fracture: Subconchoidal to uneven; brittle
Other: Alloyed to make brass and zinc die castings. Corrosion resistant coatings on steel (galvanizing) and iron. Manufacture of corrosion resitant paints, pigments and fillers etc.
Comments: Occurs in oxidized zone of zinc ore deposits

Optical Properties

Name: Smithsonite
Commodity: Zinc (Zn)
Formula: ZnCO3
Crystal System: Trigonal, hexagonal
Color:
Pleochroism:
Anisotropism:
Internal Reflections:
Reflectance 546nm:
Reflectance at 589nm:
Quant. Color Coordinates (QC):
Vickers Hardness Number (100g):
Polishing Hardness:
Cleavage: {10-11} nearly perfect
Form: Crystals rhombohedral, often curved and rough; crusts; botryoidal, stalactitic, reniform; massive, granular to compact
Alteration:
Associated Minerals:
Distinguishing Features:
Model: Occurs in oxidized zone of zinc ore deposits

Physical Properties
Name: Sperrylite
Commodity: Platinum (Pt)
Formula: PtAs2
Crystal System: Cubic
Color: Tin-white
Opacity: Opaque
Luster: Bright metallic
Streak: Black
SGLow: 10.46
SGHigh: 10.6
HardnessLow: 6
HardnessHigh: 7
Cleavage: 1
Direction: {001} indistinct
Habit: Crystals cubic or cubo-octahedral, may be highly modified
Fracture: Conchoidal; brittle
Other: Used in dentistry. High melting point and corrosion resistant . Used as electrical contacts, chemical crucibles
Comments: Associated with nickel ores in nickel deposits. Occurs in ultramafic igneous rocks and in gold-quartz veins

Optical Properties
Name: Sperrylite
Commodity: Platinum (Pt)
Formula: PtAs2
Crystal System: Cubic
Color:
Pleochroism:
Anisotropism:
Internal Reflections:
Reflectance 546nm:
Reflectance at 589nm:
Quant. Color Coordinates (QC):
Vickers Hardness Number (100g):
Polishing Hardness:
Cleavage: {001} indistinct
Form: Crystals cubic or cubo-octahedral, may be highly modified
Alteration:
Associated Minerals:
Distinguishing Features:
Model: Associated with nickel ores in nickel deposits. Occurs in ultramafic igneous rocks and in gold-quartz veins

Physical Properties

Name: Spessartine (Spessartite)

Commodity: Garnet

Formula: Mn3Al2(SiO4)3

Crystal System: Cubic

Color: Red, reddish orange, yellowish brown, reddish brown to brown

Opacity: Transparent to translucent

Luster: Vitreous

Streak: White

SGLow: 3.8

SGHigh: 4.25

HardnessLow: 7

HardnessHigh: 7.5

Cleavage: 1

Direction: {110} parting may be distinct

Habit: Crystals dodecahedrons or trapezohedrons, faces often striated; embedded grains; massive, compact

Fracture:

Other: Abrasives, gemstones

Comments: Occurs in many types of metamorphic rocks and some igneous rocks and in alluvial and beach deposits

Optical Properties

Name: Spessartite (Spessartine)

Group: Garnet

Formula: Mn3Al2(SiO4)3

Crystal System: Isometric, cubic

Color: Colorless to pale red, pale to dark brown, greenish

Form: Euhedral dodecahedrons in six sided trapezohedrons in eigth sided cross sections. Plygonal grains and aggregates.

Relief: V.high, n>balsam

Birefringence: Isotropic but may show weak birefringence

2V:

Nalpha or Nord.: 1.792-1.820

NBeta or Nextr.:

NGamma:

Optical Sign:

Orientation:

Pleochroism:

Twinning:

Cleavage: Parting parallel to {110}, irregular fractures

Extinction:

Alteration:

Features: Similar to spinel which is octohedral. Determination of RI will differentiate garnets.

Occurrence: Occurs in pegmatites and metamorphic schists and quartzites

Physical Properties

Name: Sphalerite

Commodity: Zinc (Zn)

Formula: (Zn,Fe)S

Crystal System: Cubic

Color: Brown, black, yellow, green, red, grey, white, colorless

Opacity: Transparent to translucent

Luster: Resinous to adamantine

Streak: Colorless to pale brown

SGLow: 3.9

SGHigh: 4.1

HardnessLow: 3.5

HardnessHigh: 4

Cleavage: 1

Direction: {011} perfect

Habit: Crystals tetrahedral, may appear octahedral, sometimes dodecahedral; massive, cleavable, granular; stalactitic

Fracture: Conchoidal; brittle

Other: Alloyed to make brass and zinc die castings. Corrosion resistant coatings on steel (galvanizing) and iron. Manufacture of corrosion resitant paints, pigments and fillers etc.

Comments: Associated with galena and copper sulphides in veins and in limestone replacement deposits

Optical Properties

Name: Sphalerite

Commodity: Zinc (Zn)

Formula: (Zn,Fe)S

Crystal System: Cubic

Color: Gray, sometimes with brown tint

Pleochroism: None

Anisotropism: Isotropic

Internal Reflections: Yellow brown to reddish brown

Reflectance 546nm: 16.7

Reflectance at 589nm: 16.4

Quant. Color Coordinates (QC): 0.303, 0.309, 16.6

Vickers Hardness Number (100g): 218-227, perfect

Polishing Hardness: Greater than chalcopyrite, equal to tetrahedrite, less tha pyrrhotite, magnetite

Cleavage: {011} perfect

Form: Usially anhedral aggregates. Inclusions of chalcopyrite, pyrrhotite, galena. Zoned withh light and dark bands

Alteration:

Associated Minerals: With pyrite, galena, chalcopyrite, pyrrhotite

Distinguishing Features: Internal reflections characteristic

Model: Associated with galena and copper sulphides in veins and in limestone replacement deposits

Physical Properties

Name: Spodumene

Commodity: Lithium (Li)

Formula: LiAlSi2O6

Crystal System: Monoclinic

Color: Colorless, white, grey, yellowish, greenish, emerald-green, pink, violet

Opacity: Transparent to translucent

Luster: Vitreous to dull

Streak: White

SGLow: 3.0

SGHigh: 3.2

HardnessLow: 6.5

HardnessHigh: 7.5

Cleavage: 3

Direction: {110} perfect, {010} and {100} parting

Habit: Crystals prismatic, may be flattened, striated vertically, large; massive, cleavable

Fracture:

Other: Used as a base for greases, in production of aluminum and in ceramics

Comments: Occurs in pegmatites with lepidolite, tourmaline and beryl

Optical Properties

Name: Spodumene

Group: Pyroxene

Formula: LiAlSi2O6

Crystal System: Monoclinic

Color: Colorless. Kunzite (amethystine) and hiddenite (greenish).

Form: Euhedral tabular crystals, elongated along {001}

Relief: High, n>balsam

Birefringence: Moderate, 0.013-0.027

2V: 54"-69"

Nalpha or Nord.: 1.651-1.668

NBeta or Nextr.: 1.665-1.675

NGamma: 1.677-1.681

Optical Sign: Biaxial positive

Orientation: Length slow, r<v

Pleochroism: Thick colored sections are pleochroic

Twinning: Simple twins along {100} common

Cleavage: Three perfect parallel to {110} and (11_0) at 93". arting parallel to {100}, often more distinct than cleavage.

Extinction: Oblique 23"-27" in longitudinal sections, parallel or symmetrical in cross sections

Alteration: Alters to a mixture of albite and muscovite known as cymatolite

Features: Similar to diopside, but with smaller extinction angle. Occurrence distinctive.

Occurrence: A lithium mineral found mainly in granite pegmatites with other lithium minerals

Physical Properties
Name: Stannite
Commodity: Copper (Cu)
Formula: Cu2FeSnS4
Crystal System: Tetragonal
Color: Steel-grey to greyish black, tarnishes blue
Opacity: Opaque
Luster: Metallic
Streak: Blackish
SGLow: 4.3
SGHigh: 4.5
HardnessLow: 4
HardnessHigh: 4
Cleavage: 2
Direction: {110} and {001} indistinct
Habit: Crystals pseudotetragonal or pseudododecahedral due to twinning; often striated; massive, granular
Fracture: Uneven; brittle
Other: Alloyed with various metals to produce bronzes and brasses. Used in electrical industries where high electrical and thermal conductivity is required.
Comments: Chalcopyrite is the main ore of copper. Cuprite is found in the oxidized zone of copper deposits

Optical Properties
Name: Stannite
Commodity: Copper (Cu)
Formula: Cu2FeSnS4
Crystal System: Tetragonal
Color: Brownish olive green
Pleochroism: None
Anisotropism: Isotropic
Internal Reflections: Yellow brown to redish brown
Reflectance 546nm: 26.0-27.3
Reflectance at 589nm: 26.1-27.3
Quant. Color Coordinates (QC): 0.316, 0.326, 27.1
Vickers Hardness Number (100g): 140-326
Polishing Hardness: Greater than chalcopyrite, equal to tetrahedrite, less than sphalerite
Cleavage: {110} and {001} indistinct
Form: Anhedral aggregates and with intergrowths of sphalerite, chalcopyrite and tetrahedrite. Cleavage may be visible. Polysynthetic twinning common
Alteration:
Associated Minerals: With sphalerite, chalcopyrite and tetrahedrite, With bismuth and tungsten minerals
Distinguishing Features: Pleochroism and anisotropy
Model: Chalcopyrite is the main ore of copper. Cuprite is found in the oxidized zone of copper deposits

Physical Properties

Name: Steatite (Talc)

Commodity: Talc

Formula: Mg3Si4O10(OH)2

Crystal System: Monoclinic and triclinic

Color: Pale green to dark green, greenish grey, white, silvery white, grey, brownish

Opacity: Translucent

Luster: Pearly or dull

Streak: White

SGLow: 2.58

SGHigh: 2.83

HardnessLow: 1

HardnessHigh: 1

Cleavage: 1

Direction: {001} perfect

Habit: Crystals thin tabular, to 1 cm; massive, fine-grained compact, foliated or fibrous; globular stellate groups

Fracture: Laminae flexible, inelastic

Other: Used as filler for paints, paper and rubber and in plasters, lubricants, toilet powder, chalk

Comments: Occurs in metamorphic rocks as an alteration product of olivine, pyroxene and amphibole

Optical Properties

Name: Steatite (Talc)

Group: Brittle Mica

Formula: Mg3Si4O10(OH)2

Crystal System: Monoclinic and triclinic

Color: Colorless

Form: Cleavage masses in coarse to fine platy or fibrous aggregates. Euhedral crystals unknown

Relief: Fair, n>balsam

Birefringence: Very strong, 0.030-0.050. Low, first order grey parallel to cleavage

2V: 6"=30"

Nalpha or Nord.: 1.538-1.545

NBeta or Nextr.: 1.575-1.590

NGamma: 1.575-1.590

Optical Sign: Biaxial negative

Orientation: Length slow, r>v distinct

Pleochroism:

Twinning:

Cleavage: One perfect {001}

Extinction: Parallel upto 3"

Alteration:

Features: Similar to muscovite and pyrophyllite, but physical examination of sample will usually distinguish.

Occurrence: Found within metamorphic rocks, especially schists. Alteration product of actinolite and tremolite

Physical Properties
Name: Stibnite
Commodity: Antimony (Sb)
Formula: Sb2S3
Crystal System: Orthorhombic
Color: Pale to dark lead-grey, may tarnish iridescent, bluish or blackish
Opacity: Opaque
Luster: Metallic, often brilliant
Streak: Grey to dark grey
SGLow: 4.63
SGHigh: 4.66
HardnessLow: 2
HardnessHigh: 2
Cleavage: 3
Direction: {010} perfect, easy, {100} and {110} imperfect
Habit: Crystals slender prismatic, vertically striated, often bent or twisted; bladed; columnar; massive, compact
Fracture: Subconchoidal to uneven; flexible, not elastic
Other: Flame resistant properties used in textiles and other materials. Alloying with other metals, particularly lead
Comments: Main ore of antimony. Occurs in quartz veins with galena, pyrite, realgar, orpiment and cinnabar. Occurs in epithermal and mesothermal quartz vein deposits

Optical Properties
Name: Stibnite
Commodity: Antimony (Sb)
Formula: Sb2S3
Crystal System: Orthorhombic
Color: Pale to dark lead-Gray, may tarnish iridescent, bluish or blackish
Pleochroism: Opaque
Anisotropism: Metallic, often brilliant
Internal Reflections: Gray to dark Gray
Reflectance 546nm: 2
Reflectance at 589nm: 2
Quant. Color Coordinates (QC): 4.63
Vickers Hardness Number (100g): 4.66
Polishing Hardness: 3
Cleavage: {010} perfect, easy, {100} and {110} imperfect
Form: Crystals slender prismatic, vertically striated, often bent or twisted; bladed; columnar; massive, compact
Alteration:
Associated Minerals:
Distinguishing Features: Crystals slender prismatic, vertically striated, often bent or twisted; bladed; columnar; massive, compact
Model: Main ore of antimony. Occurs in quartz veins with galena, pyrite, realgar, orpiment and cinnabar. Occurs in epithermal and mesothermal quartz vein deposits

Optical Properties

Name: Stromeyerite

Commodity: Silver

Formula: AgCuS

Crystal System: Orthorhombic

Color: Gray with violet pinkish tint

Pleochroism: Weak, Gray brown to light Gray with blue or pink tint

Anisotropism: Strong, light violet, purple, brown, orange-yellow

Internal Reflections: None

Reflectance 546nm: 26.3-29.7

Reflectance at 589nm: 25.4-28.6

Quant. Color Coordinates (QC): 0.286, 0.285, 30.1

Vickers Hardness Number (100g): 30-32, slightly fractured

Polishing Hardness: Less than galena, chalcocite

Cleavage: {001}, {110} and {111} imperfect, {111} parting

Form: In grabular aggregates and small veinlets. Intergrown with other silver minerals and Cu-Fe sulphides

Alteration:

Associated Minerals:

Distinguishing Features: Anisotropy distinctive

Model: Occurs in volcanoes and hot springs

Physical Properties

Name: Sulfur

Commodity: Sulphur

Formula: S

Crystal System: Orthorhombic

Color: Sulfur yellow to yellowish brown, yellowish grey, reddish, greenish

Opacity:

Luster:

Streak:

SGLow: 2.07

SGHigh: 2.07

HardnessLow: 1.5

HardnessHigh: 2.5

Cleavage: 4

Direction: {001}, {110} and {111} imperfect, {111} parting

Habit: Crystals bipyramidal, thick tabular or disphenoidal; massive; crusts; stalactites; powdery

Fracture: Conchoidal to uneven; brittle to somewhat sectile

Other: Production of fertilizers, sulphuric acid, insecticides, gunpowder, sulphur dioxide etc.

Comments: Occurs in volcanoes and hot springs

Physical Properties

Name: Sylvanite

Commodity: Gold (Au)

Formula: (Au,Ag)2Te4

Crystal System: Monoclinic

Color: Silver-white, yellowish tint

Opacity: Opaque

Luster: Metallic, brilliant

Streak: Silver-white

SGLow: 8.16

SGHigh: 8.16

HardnessLow: 1.5

HardnessHigh: 2

Cleavage: 1

Direction: {010} perfect

Habit: Crystals thick tabular or short prismatic, penetration twinning; bladed, columnar or granular

Fracture: Uneven; brittle

Other: Monetary use, jewellery, electronics and for decoration

Comments: Occurs in hypothermalin quartz veins associated with gold ores. Rare.

Optical Properties

Name: Sylvanite

Commodity: Tellurium (Te)

Formula: (Au,Ag)2Te4

Crystal System: Monoclinic

Color: Creamy white

Pleochroism: Distinct, creamy white to brownish

Anisotropism: Strong, light bluish Gray to dark brown

Internal Reflections: None

Reflectance 546nm: 50.3-29.0

Reflectance at 589nm: 50.8-59.0

Quant. Color Coordinates (QC):

Vickers Hardness Number (100g): 91-104, slightly fractured

Polishing Hardness: Greater than argentite, less than pyrargyrite

Cleavage: Well developed, {010} perfect

Form: Skeletal bladed crystals with well developed cleavage and polysynthetic twins

Alteration:

Associated Minerals: Greater than argentite, less than pyrarygyrite

Distinguishing Features: Crystals thick tabular or short prismatic, penetration twinning; bladed, columnar or granular

Model: Occurs in veins with gold, bismuth and lead. Tellurium is produced with selenium as a byproduct during copper refining

Physical Properties

Name: Sylvite (Sylvine)

Commodity: Potash

Formula: KCl

Crystal System: Cubic

Color: White, grey, bluish; reddish from hematite inclusions

Opacity: Transparent

Luster: Vitreous

Streak:

SGLow: 1.993

SGHigh: 1.993

HardnessLow: 2

HardnessHigh: 2

Cleavage: 1

Direction: {001} perfect

Habit: Crystals cubic; octahedrons or cubo-octahedrons rare; massive, compact to granular; crusts; columnar

Fracture: Uneven; brittle

Other: Used in fertilisers and potassium salts

Comments: Occurs in arid climates where it forms deposits in evaporite lakes. Associated with halite and carnallite

Physical Properties

Name: Talc

Commodity: Talc

Formula: Mg3Si4O10(OH)2

Crystal System: Monoclinic and triclinic

Color: Pale green to dark green, greenish grey, white, silvery white, grey, brownish

Opacity: Translucent

Luster: Pearly or dull

Streak: White

SGLow: 2.58

SGHigh: 2.83

HardnessLow: 1

HardnessHigh: 1

Cleavage: 1

Direction: {001} perfect

Habit: Crystals thin tabular, to 1 cm; massive, fine-grained compact, foliated or fibrous; globular stellate groups

Fracture: Laminae flexible, inelastic

Other: Used as filler for paints, paper and rubber and in plasters, lubricants, toilet powder, chalk

Comments: Occurs in metamorphic rocks as an alteration product of olivine, pyroxene and amphibole

Optical Properties

Name: Talc

Group: Brittle Mica

Formula: Mg3Si4O10(OH)2

Crystal System: Monoclinic and triclinic

Color: Colorless

Form: Cleavage masses in coarse to fine platy or fibrous aggregates. Euhedral crystals unknown

Relief: Fair, n>balsam

Birefringence: Very strong, 0.030-0.050. Low, first order grey parallel to cleavage

2V: 6"=30"

Nalpha or Nord.: 1.538-1.545

NBeta or Nextr.: 1.575-1.590

NGamma: 1.575-1.590

Optical Sign: Biaxial negative

Orientation: Length slow, r>v distinct

Pleochroism:

Twinning:

Cleavage: One perfect {001}

Extinction: Parallel upto 3"

Alteration:

Features: Similar to muscovite and pyrophyllite, but physical examination of sample will usually distinguish.

Occurrence: Found within metamorphic rocks, especially schists. Alteration product of actinolite and tremolite

Physical Properties
Name: Tantalite, see ferrotantalite and manganotantalite
Commodity: Niobium (Nb)
Formula: (Fe,Mn)Ta2O6
Crystal System: Orthorhombic
Color: Black to brownish black, tarnishes iridescent
Opacity: Opaque
Luster: Submetallic to weakly vitreous
Streak:
SGLow: 8.2
SGHigh: 8.2
HardnessLow: 6
HardnessHigh: 6.5
Cleavage: 2
Direction: {010} distinct, {100} less distinct
Habit: Crystals thin to thick tabular, short prismatic, equant or pyramidal; aggregates of parallel to divergent crystals
Fracture: Subconchoidal to uneven; brittle
Other: Important ferror-alloy to inhibit intergranular corrosion at high temperatures. Important metal used in electronics.
Comments: Associated with cassiterite, wolframite, spodumene and tourmaline in granite pegmatites

Optical Properties
Name: Tantalite, see ferrotantalite and manganotantalite
Commodity: Tantalum (Ta)
Formula: (Fe,Mn)Ta2O6
Crystal System: Orthorhombic
Color: Black to brownish black, tarnishes iridescent
Pleochroism: Opaque
Anisotropism: Submetallic to weakly vitreous
Internal Reflections:
Reflectance 546nm: 6
Reflectance at 589nm: 6.5
Quant. Color Coordinates (QC): 8.2
Vickers Hardness Number (100g): 8.2
Polishing Hardness: 2
Cleavage: {010} distinct, {100} less distinct
Form: Crystals thin to thick tabular, short prismatic, equant or pyramidal; aggregates of parallel to divergent crystals
Alteration:
Associated Minerals:
Distinguishing Features: Crystals thin to thick tabular, short prismatic, equant or pyramidal; aggregates of parallel to divergent crystals
Model: Associated with cassiterite, wolframite, spodumene and tourmaline in granite pegmatites. Tantalum is also recovered from tin slags as a byproduct

Physical Properties

Name: Tennantite

Commodity: Copper (Cu)

Formula: $(Cu,Fe)_{12}As_4S_{13}$

Crystal System: Cubic

Color: Steel-grey to iron-black

Opacity: Opaque

Luster: Metallic, sometimes splendent

Streak: Black to brown to dark red

SGLow: 4.59

SGHigh: 4.75

HardnessLow: 3

HardnessHigh: 4.5

Cleavage: 0

Direction:

Habit: Crystals tetrahedral, often modified, twinned; massive, coarse granular to compact

Fracture: Subconchoidal to uneven; brittle

Other: Alloyed with various metals to produce bronzes and brasses. Used in electrical industries where high electrical and thermal conductivity is required.

Comments: Chalcopyrite is the main ore of copper. . Tennantite and Tetrahedrite are found in veins with silver, lead, copper and zinc ores

Optical Properties

Name: Tennantite

Commodity: Copper (Cu)

Formula: $(Cu,Fe)_{12}As_4S_{13}$

Crystal System: Cubic

Color: Gray, sometimes with greenish or bluish tint

Pleochroism: None

Anisotropism: Isotropic

Internal Reflections: Common, reddish

Reflectance 546nm: 30.0

Reflectance at 589nm: 29.8

Quant. Color Coordinates (QC): 0.304, 0.310, 30.0

Vickers Hardness Number (100g): 297-354, perfect

Polishing Hardness: Greater than galena, Equal to chalcopyrite, less than sphalerite

Cleavage:

Form: Aggregates of anhedral grains. Zoning may be visible

Alteration:

Associated Minerals: With Cu, Fe sulphides, sphalerite, galena, arsenopyrite and sulfosalts

Distinguishing Features: Reddish internal reflections and color

Model: Chalcopyrite is the main ore of copper. . Tennantite and Tetrahedrite are found in veins with silver, lead, copper and zinc ores

Optical Properties

Name: Tenorite

Commodity: Thorium (Th)

Formula: CuO

Crystal System: Monoclinic

Color: Gray to Gray white

Pleochroism: Strong, Gray to Gray white

Anisotropism: Strong, blue to Gray

Internal Reflections: None

Reflectance 546nm: 20.9-25.2

Reflectance at 589nm:

Quant. Color Coordinates (QC):

Vickers Hardness Number (100g): 304-339

Polishing Hardness: Greater than chalcocite, less than goethite, cuprite

Cleavage: {001} poor

Form: Aggregates of acicular crystals and concentric aggregates. May show lamellar twinning

Alteration:

Associated Minerals: With Cu and Fe oxides

Distinguishing Features: Pleochroism and anisotropy distinctive

Model: With copper deposits

Physical Properties

Name: Tetrahedrite

Commodity: Copper (Cu)

Formula: $(Cu,Fe)12Sb4S13$

Crystal System: Cubic

Color: Steel-grey to iron-black

Opacity: Opaque

Luster: Metallic, sometimes splendent

Streak: Black to brown to dark red

SGLow: 4.6

SGHigh: 5.1

HardnessLow: 3

HardnessHigh: 4.5

Cleavage: 0

Direction:

Habit: Crystals tetrahedral, modified, contact penetration twins; massive, coarse granular to compact

Fracture: Subconchoidal to uneven; brittle

Other: Alloyed with various metals to produce bronzes and brasses. Used in electrical industries where high electrical and thermal conductivity is required.

Comments: Chalcopyrite is the main ore of copper. . Tennantite and Tetrahedrite are found in veins with silver, lead, copper and zinc ores

Optical Properties

Name: Tetrahedrite

Commodity: Copper (Cu)

Formula: $(Cu,Fe)12Sb4S13$

Crystal System: Cubic

Color: Gray with olive or brownish tint

Pleochroism: None

Anisotropism: Isotropic

Internal Reflections: Reddish, but rare

Reflectance 546nm: 30.5

Reflectance at 589nm: 30.0

Quant. Color Coordinates (QC): 0.305, 0.314, 30.2

Vickers Hardness Number (100g): 285-322, slightly fractured

Polishing Hardness: Greater than galena, equal to chalcopyrite, less than sphalerite

Cleavage:

Form: Anhedral aggregates, may be zoned

Alteration: Solid solution with tennantite

Associated Minerals: With other Cu-Fe sulphides

Distinguishing Features: Prismatic crystals, color and isotropism

Model: Chalcopyrite is the main ore of copper. . Tennantite and Tetrahedrite are found in veins with silver, lead, copper and zinc ores

Physical Properties

Name: Thorianite

Commodity: Thorium (Th)

Formula: ThO_2

Crystal System: Cubic

Color: Brownish yellow, yellow to orange-yellow, greenish black to green

Opacity: Translucent to opaque

Luster: Vitreous, resinous to greasy

Streak:

SGLow: 9.99

SGHigh: 9.99

HardnessLow: 6.5

HardnessHigh: 7

Cleavage: 1

Direction: {001} poor

Habit: Crystals cubic or cubo-octahedral; embedded crystals or grains

Fracture: Conchoidal to uneven; brittle

Other: Radioactive. Used to produce thorium alloys, gas mantles, in electrical instruments and in medicine

Comments: Occurs in some heavy mineral strandline beach deposits

Physical Properties

Name: Tin

Commodity: Tin (Sn)

Formula: Sn

Crystal System: Tetragonal

Color: Tin-white

Opacity: Opaque

Luster: Metallic

Streak:

SGLow: 7.31

SGHigh: 7.31

HardnessLow: 1.5

HardnessHigh: 2

Cleavage: 0

Direction:

Habit: Embedded grains, minute; rounded grains in placers

Fracture: Hackly; ductile, malleable

Other: Alloyed to produce solder, pewter,, bronze, type-metal. Used as tin plate for cans and to produce casting metal

Comments: Occurs in granites and pegmatites with cassiterite. Also forms alluvial deposits

Optical Properties

Name: Tin

Commodity: Tin (Sn)

Formula: Sn

Crystal System: Tetragonal

Color: Tin-white

Pleochroism: Opaque

Anisotropism: Metallic

Internal Reflections:

Reflectance 546nm: 1.5

Reflectance at 589nm: 2

Quant. Color Coordinates (QC): 7.31

Vickers Hardness Number (100g): 7.31

Polishing Hardness: 0

Cleavage:

Form: Embedded grains, minute; rounded grains in placers

Alteration:

Associated Minerals:

Distinguishing Features: Embedded grains, minute; rounded grains in placers

Model: Occurs in granites and pegmatites with cassiterite. Also forms alluvial deposits

Physical Properties
Name: Titanite
Commodity: Titanium (Ti)
Formula: CaTiSiO5
Crystal System: Monoclinic
Color: Colorless, yellow, green, grey, brown, rose-red, black, pinkish red
Opacity: Transparent to nearly opaque
Luster: Adamantine to resinous
Streak: White
SGLow: 3.45
SGHigh: 3.55
HardnessLow: 5
HardnessHigh: 5.5
Cleavage: 2
Direction: {110} distinct, {221} parting
Habit: Crystals wedge-shaped or prismatic, twinned; massive, compact or lamellar
Fracture:
Other: High strength and corrosion resistance. Mostly used in aircraft and aerospace industries. Used in chemical and desalination plants and heatbexchanger tubing in power plants
Comments: Associated with monazite, zircon, ilmenite etc in alluvial sands and heavy mineral strandline beach deposits. An accessory mineral in igneous rocks concentrated by sedimentary processes

Optical Properties
Name: Titanite
Commodity: Titanium (Ti)
Formula: CaTiSiO5
Crystal System: Monoclinic
Color: Colorless, yellow, green, Gray, brown, rose-red, black, pinkish red
Pleochroism: Transparent to nearly opaque
Anisotropism: Adamantine to resinous
Internal Reflections: White
Reflectance 546nm: 5
Reflectance at 589nm: 5.5
Quant. Color Coordinates (QC): 3.45
Vickers Hardness Number (100g): 3.55
Polishing Hardness: 2
Cleavage: {110} distinct, {221} parting
Form: Crystals wedge-shaped or prismatic, twinned; massive, compact or lamellar
Alteration:
Associated Minerals:
Distinguishing Features: Crystals wedge-shaped or prismatic, twinned; massive, compact or lamellar
Model: Associated with monazite, zircon, ilmenite etc in alluvial sands and heavy mineral strandline beach deposits. An accessory mineral in igneous rocks concentrated by sedimentary processes

Physical Properties
Name: Torbernite
Commodity: Uranium (U)
Formula: Cu(UO2)2(PO4)2+8-12H2O
Crystal System: Tetragonal
Color: Green
Opacity: Transparent to translucent
Luster: Vitreous to subadamantine
Streak: Lighter than color
SGLow: 3.22
SGHigh: 3.22
HardnessLow: 2
HardnessHigh: 2.5
Cleavage: 2
Direction: {001} perfect, {100} indistinct
Habit: Crystals thin to thick tabular, rectangular or octagonal shape, rarely pyramidal; scaly or lamellar aggregates; massive
Fracture: Brittle
Other: Radioactve, used as nuclear fuel
Comments: Occurs in the oxidized zone of uranium deposits with autunite and carnotite

Optical Properties
Name: Torbernite
Commodity: Uranium (U)
Formula: Cu(UO2)2(PO4)2+8-12H2O
Crystal System: Tetragonal
Color: Green
Pleochroism: Transparent to translucent
Anisotropism: Vitreous to subadamantine
Internal Reflections: Lighter than color
Reflectance 546nm: 2
Reflectance at 589nm: 2.5
Quant. Color Coordinates (QC): 3.22
Vickers Hardness Number (100g): 3.22
Polishing Hardness: 2
Cleavage: {001} perfect, {100} indistinct
Form: Crystals thin to thick tabular, rectangular or octagonal shape, rarely pyramidal; scaly or lamellar aggregates; massive
Alteration:
Associated Minerals:
Distinguishing Features: Crystals thin to thick tabular, rectangular or octagonal shape, rarely pyramidal; scaly or lamellar aggregates; massive
Model: Occurs in the oxidized zone of uranium deposits with autunite and carnotite

Physical Properties

Name: Tourmaline (Elbaite)

Commodity: Lithium (Li)

Formula: Na(Li,Al)3Al6(BO3)3Si6O18(OH)4

Crystal System: Trigonal

Color: Green, blue, red, yellow, white, colorless

Opacity: Transparent to translucent

Luster: Vitreous

Streak: Colorless

SGLow: 3.03

SGHigh: 3.10

HardnessLow: 7

HardnessHigh: 7

Cleavage: 2

Direction: {11-20} and {10-11} indistinct

Habit: Crystals short to long prismatic, vertically striated, hemimorphic; massive, compact, columnar to fibrous

Fracture: Conchoidal to uneven; brittle

Other: Used as a base for greases, in production of aluminum and in ceramics

Comments: Occurs in pegmatites with lepidolite, tourmalinae, beryl and other lithium minerals. Not a commercial source of lithium.

Optical Properties

Name: Tourmaline (Elbaite)

Commodity: Lithium (Li)

Formula: Na(Li,Al)3Al6(BO3)3Si6O18(OH)4

Crystal System: Trigonal

Color: Green, blue, red, yellow, white, colorless

Pleochroism: Transparent to translucent

Anisotropism: Vitreous

Internal Reflections: Colorless

Reflectance 546nm: 7

Reflectance at 589nm: 7

Quant. Color Coordinates (QC): 3.03

Vickers Hardness Number (100g): 3.10

Polishing Hardness: 2

Cleavage: {11-20} and {10-11} indistinct

Form: Crystals short to long prismatic, vertically striated, hemimorphic; massive, compact, columnar to fibrous

Alteration:

Associated Minerals:

Distinguishing Features: Crystals short to long prismatic, vertically striated, hemimorphic; massive, compact, columnar to fibrous

Model: Occurs in pegmatites with lepidolite, tourmalinae, beryl and other lithium minerals. Not a commercial source of lithium.

Physical Properties

Name: Ulexite

Commodity: Borates

Formula: NaCaB5O6(OH)6+5H2O

Crystal System: Triclinic

Color: Crystals colorless, aggregates white

Opacity:

Luster: Vitreous on crystals, aggregates satiny

Streak:

SGLow: 1.96

SGHigh: 1.96

HardnessLow: 2.5

HardnessHigh: 2.5

Cleavage: 3

Direction: {010} perfect, {1-10} good, {110} poor

Habit: Crystals acicular, greatly elongated, in nodules and lens-like masses; crusts and compact veins

Fracture:

Other: Used in manufacture of insulating fibreglass, as fluxes for manufacture of glass and enamels. Borax is used in cloth manufacture and tanning, soap and glue, as preservatives and antiseptics

Comments: Precipitated in evaporite basins in arid climates. Associated with borax

Optical Properties

Name: Ulexite

Commodity: Borates

Formula: NaCaB5O6(OH)6+5H2O

Crystal System: Triclinic

Color: Crystals colorless, aggregates white

Pleochroism:

Anisotropism: Vitreous on crystals, aggregates satiny

Internal Reflections:

Reflectance 546nm: 2.5

Reflectance at 589nm: 2.5

Quant. Color Coordinates (QC): 1.96

Vickers Hardness Number (100g): 1.96

Polishing Hardness: 3

Cleavage: {010} perfect, {1-10} good, {110} poor

Form: Crystals acicular, greatly elongated, in nodules and lens-like masses; crusts and compact veins

Alteration:

Associated Minerals:

Distinguishing Features: Crystals acicular, greatly elongated, in nodules and lens-like masses; crusts and compact veins

Model: Precipitated in evaporite basins in arid climates. Associated with borax

Physical Properties
Name: Ulvospinel
Commodity: Titanium (Ti)
Formula: TiFe2O4
Crystal System: Cubic
Color: Iron-black
Opacity:
Luster:
Streak:
SGLow: 4.77
SGHigh: 4.77
HardnessLow:
HardnessHigh:
Cleavage:
Direction:
Habit: Massive, microscopic exsolution lamellae
Fracture:
Other: High strength and corrosion resistance. Mostly used in aircraft and aerospace industries. Used in chemical and desalination plants and heatbexchanger tubing in power plants
Comments: Associated with monazite, zircon, ilmenite etc in alluvial sands and heavy mineral strandline beach deposits. An accessory mineral in igneous rocks concentrated by sedimentary processes

Optical Properties
Name: Ulvospinel
Commodity: Titanium (Ti)
Formula: TiFe2O4
Crystal System: Cubic
Color: Brown to reddish brown
Pleochroism: Not present
Anisotropism: Isotropic
Internal Reflections: None
Reflectance 546nm: 15.5
Reflectance at 589nm: 16.35
Quant. Color Coordinates (QC):
Vickers Hardness Number (100g):
Polishing Hardness: Greater than magnetite
Cleavage:
Form: octohedral crystals occassionally. Usually as exsolution lamellae in Ti magnetite, giving a cloth weave texture. Also as matrix containing oriented cubes of magnetite
Alteration:
Associated Minerals: With ilmenite and magnetite
Distinguishing Features: Color, isotropism, occurrence
Model: Associated with monazite, zircon, ilmenite etc in alluvial sands and heavy mineral strandline beach deposits. An accessory mineral in igneous rocks concentrated by sedimentary processes

Physical Properties

Name: Uraninite (Pitchblende)

Commodity: Uranium (U)

Formula: UO_2

Crystal System: Cubic

Color: Black to brownish or greyish black

Opacity: Opaque

Luster: Submetallic, pitchy, greasy

Streak: Black to brownish black, greyish

SGLow: 7.5

SGHigh: 10.0

HardnessLow: 5

HardnessHigh: 6

Cleavage: 1

Direction: Octahedral

Habit: Crystals cubic, cubo-octahedral or modified octahedrons or dodecahedrons; massive, dense to granular; botryoidal

Fracture: Conchoidal to uneven; brittle

Other: Radioactve, used as nuclear fuel

Comments: Occurs in veins withtin, copper, lead and arsenic sulphides. Associated with vanadium and copper ores in "roll type" sandstone deposits. Also associated with gold in ancient conglomerates such as Elliot Lake and the Witwatersrand

Optical Properties

Name: Uraninite (Pitchblende)

Commodity: Uranium (U)

Formula: UO_2

Crystal System: Cubic

Color: Brownish Gray

Pleochroism: None

Anisotropism: Isotropic

Internal Reflections: Dark brown to reddish brown

Reflectance 546nm: 13.6

Reflectance at 589nm: 13.6

Quant. Color Coordinates (QC): 0.305, 0.308, 13.7

Vickers Hardness Number (100g): 499-548, slightly fractured, 50g

Polishing Hardness: Greater than magnetite, less than pyrite

Cleavage: Often visible {100} and {111}

Form: Zoned crystals and colloform, oollitic and denddritic aggregates. Twinning common

Alteration:

Associated Minerals: With Cu, Fe sulphides and other uranium minerals, may contain inclusions of gold

Distinguishing Features: Internal reflections distinctive

Model: Occurs in veins withtin, copper, lead and arsenic sulphides. Associated with vanadium and copper ores in "roll type" sandstone deposits. Also associated with gold in ancient conglomerates such as Elliot Lake and the Witwatersrand

Physical Properties

Name: Uvarovite

Commodity: Garnet

Formula: Ca3Cr2(SiO4)3

Crystal System: Cubic

Color: Emerald-green

Opacity: Transparent to translucent

Luster: Vitreous

Streak: White

SGLow: 3.4

SGHigh: 3.8

HardnessLow: 6.5

HardnessHigh: 7

Cleavage: 0

Direction:

Habit: Crystals dodecahedrons or trapezohedrons; massive, coarse granular; embedded grains

Fracture:

Other: Abrasives, gemstones

Comments: Occurs in many types of metamorphic rocks and some igneous rocks and in alluvial and beach deposits

Optical Properties

Name: Uvarovite (Garnet)

Group: Garnet

Formula: Ca2Cr2(SiO4)3

Crystal System: Isometric, cubic

Color: Colorless to pale red, pale to dark brown, often zoned, greenish

Form: Euhedral dodecahedrons in six sided trapezohedrons in eigth sided cross sections. Plygonal grains and aggregates.

Relief: V.high, n>balsam

Birefringence: Isotropic, but may show weak birefringence

2V:

Nalpha or Nord.: 1.838-1.870

NBeta or Nextr.:

NGamma:

Optical Sign:

Orientation:

Pleochroism:

Twinning:

Cleavage: Parting parallel to {110}, irregular fractures

Extinction:

Alteration:

Features: Similar to spinel, which is octohedral. Determination of RI will differentiate garnets.

Occurrence: Rare, secondary in chromite and in some contact metamorphic zones.

Optical Properties

Name: Vallerite

Commodity: Copper

Formula: (Fe,Cu)S2.(Mg,Al)(OH)2

Crystal System: Hexagonal

Color: Bronze to Gray

Pleochroism: Very strong, bronze to Gray

Anisotropism: Extreme, white to Gray bronze with satin like texture

Internal Reflections: None

Reflectance 546nm: 14.2

Reflectance at 589nm: 22.0

Quant. Color Coordinates (QC):

Vickers Hardness Number (100g):

Polishing Hardness: Greater than chalcopyrite, equal to cubanite, less than pyrrhotite

Cleavage:

Form: Occurs as veinlets, interstitial fillings and inclusions in and around chalcopyrite, pyrrhotite, pentlandite, magnetite

Alteration:

Associated Minerals: interstitial fillings and inclusions in and around chalcopyrite, pyrrhotite, pentlandite, magnetite

Distinguishing Features: Anisotropy, pleochroism distinctive

Model: Occurs in many types of metamorphic rocks and some igneous rocks and in alluvial and beach deposits

Physical Properties

Name: Vanadinite

Commodity: Vanadium (V)

Formula: Pb5(VO4)3Cl

Crystal System: Hexagonal

Color: Bright red, orange-red, brownish, pale yellowish to brownish yellow, white, colorless

Opacity: Transparent to nearly opaque

Luster: Resinous to subadamantine

Streak: White or yellowish

SGLow: 6.88

SGHigh: 6.88

HardnessLow: 3

HardnessHigh: 3

Cleavage:

Direction:

Habit: Crystals short to long prismatic, acicular to hair-like, may be cavernous or skeletal; globular

Fracture: Conchoidal to uneven; brittle

Other: Alloyed with steel to produce high strength steels including tool steels, oil and gas pipelines, structural steels. Used in chemical and oil industries as catalysts and glass colouring agent

Comments: Occurs in oxidized zone of lead and lead zinc deposits

Optical Properties

Name: Violorite

Commodity: Nickel (Ni)

Formula: FeNi2S4

Crystal System: Hexagonal

Color: Brownish Gray with violet tint

Pleochroism: None

Anisotropism: Isotropic

Internal Reflections: None

Reflectance 546nm: 46.6

Reflectance at 589nm:

Quant. Color Coordinates (QC):

Vickers Hardness Number (100g): 241-373

Polishing Hardness: Greater than chalcopyrite, sphalerite, equal to pentlandite, less than pyrrhotite

Cleavage:

Form: Equant anhedral grains and fine lamellar intergrowths with millerite and chalcopyrite

Alteration: Alters along grain boundaries and fractures of pentlandite, pyrhhotite and millerite.

Associated Minerals: Pyrrhotite, pentlandite, millerite and chalcopyrite

Distinguishing Features: Color, isotropism and occurrence

Model: Occurs in oxidized zone of lead and lead zinc deposits

Physical Properties
Name: Vermiculite
Commodity: Vermiculite
Formula: (Mg,Fe,Al)3(Al,Si)4O10(OH)2+4H2O
Crystal System: Monoclinic
Color:
Opacity:
Luster:
Streak:
SGLow:
SGHigh:
HardnessLow:
HardnessHigh:
Cleavage:
Direction:
Habit:
Fracture:
Other: Excellent thermal and sound insulating properties, fire resistant, light and inert. Used for building materials
Comments: An alteration product of magnesian micas in carbonatites

Optical Properties
Name: Vermiculite
Group: Vermiculite, Clay
Formula: (Mg,Fe,Al)3(Al,Si)4O10(OH)2+4H2O
Crystal System: Monoclinic
Color:
Form:
Relief:
Birefringence:
2V:
Nalpha or Nord.:
NBeta or Nextr.:
NGamma:
Optical Sign:
Orientation:
Pleochroism:
Twinning:
Cleavage:
Extinction:
Alteration:
Features: A group of silicates with the general formula above. A group of clay minerals derived from alteration of mica
Occurrence: A weathering product of igneous and metamorphic rocks and in clay beds. Also common as a diagenetic clay mineral in sedimentary beds

Physical Properties

Name: Wavellite

Commodity: Phosphates

Formula: Al3(PO4)2(OH,F)3+5H2O

Crystal System: Orthorhombic

Color: White to green, yellowish green to brownish black

Opacity: Transparent to translucent

Luster: Vitreous to resinous or pearly

Streak:

SGLow: 4

SGHigh: 2.36

HardnessLow: White

HardnessHigh: 3.5

Cleavage: 2.36

Direction: 3

Habit: Subconchoidal to uneven; brittle

Fracture: {110} perfect, {101} good, {010} distinct

Other: Fertiliser

Comments: Formed by precipitation from sea water, replacement of limestones or shelly marine sediments, in guano deposits and accumulation of grains from pre-existing rocks

Physical Properties
Name: Willemite
Commodity: Zinc (Zn)
Formula: Zn2SiO4
Crystal System: Trigonal
Color: Colorless, white, green, yellow, red, brown, grey
Opacity: Transparent to translucent
Luster: Vitreous to resinous
Streak: Colorless
SGLow: 3.89
SGHigh: 4.19
HardnessLow: 5.5
HardnessHigh: 5.5
Cleavage: 2
Direction: {0001} and {11-20} poor
Habit: Crystals stout to long slender prismatic, hexagonal; disseminated grains; massive, compact or fibrous
Fracture: Conchoidal to uneven; brittle
Other: Alloyed to make brass and zinc die castings. Corrosion resistant coatings on steel (galvanizing) and iron. Manufacture of corrosion resitant paints, pigments and fillers etc.
Comments: Associated with zincite and franklinite in contact metamorposed limestones

Optical Properties
Name: Willemite
Commodity: Zinc (Zn)
Formula: Zn2SiO4
Crystal System: Trigonal
Color: Colorless, white, green, yellow, red, brown, Gray
Pleochroism: Transparent to translucent
Anisotropism: Vitreous to resinous
Internal Reflections: Colorless
Reflectance 546nm: 5.5
Reflectance at 589nm: 5.5
Quant. Color Coordinates (QC): 3.89
Vickers Hardness Number (100g): 4.19
Polishing Hardness: 2
Cleavage: {0001} and {11-20} poor
Form: Crystals stout to long slender prismatic, hexagonal; disseminated grains; massive, compact or fibrous
Alteration:
Associated Minerals:
Distinguishing Features: Crystals stout to long slender prismatic, hexagonal; disseminated grains; massive, compact or fibrous
Model: Associated with zincite and franklinite in contact metamorposed limestones

Physical Properties

Name: Witherite

Commodity: Barium (Ba)

Formula: BaCO3

Crystal System: Orthorhombic

Color: Colorless, white, grey, may have pale yellow, greenish or brown tint

Opacity:

Luster:

Streak:

SGLow: 4.29

SGHigh: 4.29

HardnessLow: 3

HardnessHigh: 3.5

Cleavage: 2

Direction: {010} distinct, {110} imperfect

Habit: Crystals pseudohexagonal dipyramids formed by twinning, or tabular or short prismatic, with convex base; massive, granular

Fracture: Uneven

Other:

Comments: Occurs in hydrothermal veins with galena, anglesite and baryte

Optical Properties

Name: Witherite

Commodity: Barium (Ba)

Formula: BaCO3

Crystal System: Orthorhombic

Color: Colorless, white, Gray, may have pale yellow, greenish or brown tint

Pleochroism:

Anisotropism:

Internal Reflections:

Reflectance 546nm: 3

Reflectance at 589nm: 3.5

Quant. Color Coordinates (QC): 4.29

Vickers Hardness Number (100g): 4.29

Polishing Hardness: 2

Cleavage: {010} distinct, {110} imperfect

Form: Crystals pseudohexagonal dipyramids formed by twinning, or tabular or short prismatic, with convex base; massive, granular

Alteration:

Associated Minerals:

Distinguishing Features: Crystals pseudohexagonal dipyramids formed by twinning, or tabular or short prismatic, with convex base; massive, granular

Model: Occurs in hydrothermal veins with galena, anglesite and baryte

Physical Properties

Name: Wolframite (Wolfram)

Commodity: Tungsten (W)

Formula: (Fe,Mn)WO4

Crystal System: Monoclinic

Color: Reddish-brown, black

Opacity: Transparent to nearly opaque

Luster: Submetallic to resinous

Streak: Reddish-brown to black

SGLow: 7.12

SGHigh: 7.51

HardnessLow: 4

HardnessHigh: 4.5

Cleavage: 1

Direction: {010} perfect

Habit: Prismatic, striated crystals, granular, massive or bladed

Fracture: Uneven

Other: High wear resistance. Tungsten carbide for cutting and drilling tools. Used in lamp filaments, electronic parts, electrical contacts. Important steel alloy to produce tool steels

Comments: Associated with molybdenite and cassiterite in granites and pegmatites. Occurs in contact metamorphic aureoles around granite plutons with scheelite

Optical Properties

Name: Wolframite (Wolfram)

Commodity: Tungsten (W)

Formula: (Fe,Mn)WO4

Crystal System: Monoclinic

Color: Gray to white in air, Gray with brown or yellow tint in oil

Pleochroism: Weak

Anisotropism: Weak to distinct, yellow to Gray

Internal Reflections: Deep red

Reflectance 546nm: 15.0-16.2

Reflectance at 589nm: 14.7-15.9

Quant. Color Coordinates (QC):

Vickers Hardness Number (100g): 312-342, slightly fractured

Polishing Hardness: Greater than magnetite, scheelite, less than pyrite, arsenopyrite

Cleavage: Distinct {010} perfect

Form: Euhedral plates and interpenetrating laths. Often twinned

Alteration:

Associated Minerals: With scheelite, arsenopyrite, chalcopyrite, molybdenite, bismuth, gold and cassiterite

Distinguishing Features: Anisotropy and internal reflections

Model: Associated with molybdenite and cassiterite in granites and pegmatites. Occurs in contact metamorphic aureoles around granite plutons with scheelite

Physical Properties

Name: Wulfenite

Commodity: Molybdenum (Mo)

Formula: $PbMoO_4$

Crystal System: Tetragonal

Color: Yellow to orange, orange-brown to brown, yellowish grey, tan, pink, blue

Opacity: Transparent to translucent

Luster: Resinous to adamantine

Streak: White

SGLow: 6.5

SGHigh: 7.0

HardnessLow: 2.75

HardnessHigh: 3

Cleavage: 3

Direction: {011} distinct, {001} and {013} indistinct

Habit: Crystals square tabular, often thin, or octahedral, prismatic or cuboidal; massive, coarse to fine granular; twinning uncommon

Fracture: Subconchoidal to uneven; brittle

Other: A ferro-alloy. Used in manufacture of electrodes and furnace parts and as a catalyst, corrosion inhibitor and additive to lubricants

Comments: Occurs in oxidized zone of lead and molybdenum ore deposits with anglesite, cerrusite and vanadinite

Physical Properties

Name: Xenotime-[Y]

Commodity: Rare Earths, Yttrium (Y)

Formula: YPO4

Crystal System: Tetragonal

Color: Yellowish brown to reddish brown, pale grey, pale yellow, greenish, reddish

Opacity: Translucent to opaque

Luster: Vitreous to resinous

Streak:

SGLow: 4.4

SGHigh: 5.1

HardnessLow: 4

HardnessHigh: 5

Cleavage: 1

Direction: {100} perfect

Habit: Crystals short to long prismatic, pyramidal or equant; rosettes; aggregates of rough crystals

Fracture: Splintery to uneven; brittle

Other: Cerium subgroup is most important of rare earths. Used as a catalyst in petroleum refining, iron-cerium alloys in lighter flints and used in ceramic and glass industries and for manufacture of color televisions

Comments: Associated with monazite, rutile, zircon and ilmenite in heavy mineral strandline beach deposits. Also in pegmatites and alluvial deposits

Optical Properties

Name: Xenotime-[Y]

Commodity: Rare Earths, Yttrium (Y)

Formula: YPO4

Crystal System: Tetragonal

Color:

Pleochroism:

Anisotropism:

Internal Reflections:

Reflectance 546nm:

Reflectance at 589nm:

Quant. Color Coordinates (QC):

Vickers Hardness Number (100g):

Polishing Hardness:

Cleavage: {100} perfect

Form: Crystals short to long prismatic, pyramidal or equant; rosettes; aggregates of rough crystals

Alteration:

Associated Minerals:

Distinguishing Features:

Model: Associated with monazite, rutile, zircon and ilmenite in heavy mineral strandline beach deposits. Also in pegmatites and alluvial deposits

Physical Properties

Name: Zincite

Commodity: Zinc (Zn)

Formula: (Zn,Mn)O

Crystal System: Hexagonal

Color: Deep yellow to orange-yellow to dark red

Opacity: Transparent to translucent

Luster: Subadamantine

Streak: Orange-yellow

SGLow: 5.68

SGHigh: 5.68

HardnessLow: 4

HardnessHigh: 4

Cleavage: 2

Direction: {10-10} perfect; {0001} parting

Habit: Crystals hemimorphic pyramidal, often corroded and rounded; massive, platy, compact or granular

Fracture: Conchoidal; brittle

Other: Alloyed to make brass and zinc die castings. Corrosion resistant coatings on steel (galvanizing) and iron. Manufacture of corrosion resitant paints, pigments and fillers etc.

Comments: Associated with franklinite and willemite in contact metamorposed limestones

Optical Properties

Name: Zincite

Commodity: Zinc (Zn)

Formula: (Zn,Mn)O

Crystal System: Hexagonal

Color: Pinkish brown

Pleochroism: Masked by internal reflections

Anisotropism: Masked by internal reflections

Internal Reflections: Red to yellowish

Reflectance 546nm: 11.8

Reflectance at 589nm: 11.4-11.5

Quant. Color Coordinates (QC):

Vickers Hardness Number (100g): 190-219, slightly fractured

Polishing Hardness: Less than franklinite, hausmannite

Cleavage: Often visible {10-10} perfect; {0001} parting

Form: Usually anhedral rounded grains

Alteration:

Associated Minerals: With franklinite

Distinguishing Features: Color and internal reflections distinctive

Model: Associated with franklinite and willemite in contact metamorposed limestones

Physical Properties

Name: Zircon

Commodity: Zirconium (Zr)

Formula: ZrSiO4

Crystal System: Tetragonal

Color: Colorless, brown, green, grey, yellow, red

Opacity: Transparent

Luster: Vitreous to adamantine

Streak:

SGLow: 4.6

SGHigh: 4.7

HardnessLow: 7.5

HardnessHigh: 7.5

Cleavage: 2

Direction: {110} imperfect; {111} poor

Habit: Crystals short to long prismatic, or dipyramidal; sheaf-like or radial-fibrous aggregates; irregular grains

Fracture: Uneven; conchoidal when metamict; brittle

Other: Corosion resistant. Alloyed with iron, silicon and tungsten for use in nuclear reactors and for removing oxides and nitrides from steels. Used in chemical plant equipment

Comments: Associated with monazite, ilmenite etc in alluvial sands and heavy mineral strandline beach deposits. An accessory mineral in igneous rocks concentrated by sedimentary processes

Optical Properties

Name: Zircon

Group: Garnet

Formula: ZrSiO4

Crystal System: Tetragonal

Color: Colorless to pale colors

Form: Minute, short prismatic crystals, surrounded by pleochroic haloes

Relief: V.high, n>balsam

Birefringence: Very strong, 0.060-0.062

2V:

Nalpha or Nord.: 1.925-1.931

NBeta or Nextr.: 1.985-1.993

NGamma:

Optical Sign: Uniaxial positive

Orientation: Length slow

Pleochroism:

Twinning:

Cleavage: Two {110} imperfect; {111} poor, not usually visible

Extinction: Parallel

Alteration: Metamict alteration to crytolite, an amorphous mineraloid

Features: High relief and cross section distinctive

Occurrence: Widespread mineral in igneous rocks, especially granites where it is mostly an accessory mineral. Widespread detrital mineral

Alphabetical Index of Economic Minerals

Alabandite	Calaverite
Albite	Calcite
Allargentum	Carnallite
Almandine	Carnallite
Alunite	Carnotite
Amblygonite	Carnotite
Amosite (fibrous form of ferro-gedrite) (Grunerite)	Carrollite
Andalusite	Cassiterite
Andradite	Celestite
Anglesite	Cerussite
Anhydrite	Chalcocite
Anorthite	Chalcopyrite
Antimony	Chlorargyrite (Cerargyrite)
Apatite (Group Name)	Chromite
Argentite	Chromite
Argyrodite	Chrysocolla
Arsenic	Chrysotile (subgroup name of Serpintinites)
Arsenopyrite	Cinnabar
Autunite	Clausthalite
Azurite	Cobaltite
Baddeleyite	Colemanite
Baryte	Coltan (Columbite-Tantalite)
Bastnaesite-[Ce]	Columbite – Tantalite
Bastnaesite-[Ce]	Copper
Bastnaesite-[La]	Corundum
Bastnaesite-[Y]	Covellite
Bauxite (Group Name)	Crocidolite (fibrous variety of Riebeckite)
Berthierite	Cryolite
Beryl	Cuprite
Berzelianite	Diamond
Biotite	Diaspore
Bismuth	Dolomite
Bismuthinite	Dolomite
Bismutite	Enargite
Bixbyite	Epsomite
Bohmite (Boehmite)	Eucairite
Boracite	Ferrotantalite
Borax	Fluorapatite
Bornite	Fluorite
Boulangerite	Franklinite
Bournonite	Galena
Braunite	Galena
Breithauptite	Gallium
Brucite	Gibbsite
Calamine (Hemimorphite)	Goethite

Alphabetical Index of Economic Minerals cont...

Gold	Niccolite (Nickeline)
Goldamalgam	Niter
Graphite	Orpiment
Greenockite	Orthoclase
Grossular	Osmiridium
Grunerite	Patronite
Gypsum	Pentlandite
Hafnium	Phlogopite
Halite	Phosphate Rock
Hematite	Plagioclase
Hemimorphite	Platinum
Ilmenite	Pollucite
Ilmenite	Proustite
Indium	Psilomelane
Jamesonite	Pyrargyrite
Kaolinite	Pyrite
Kernite	Pyrite
Kyanite	Pyrochlore
Lepidolite	Pyrochlore
Lepidolite	Pyrolusite
Lepidolite	Pyrope
Limonite	Pyrrhotite
Linnaeite	Quartz (alpha-Quartz)
Magnesite	Rammelsbergite
Magnesite	Realgar
Magnetite	Roscoelite
Malachite	Rutile
Manganite	Rutile
Manganotantalite	Safflorite
Marcasite	Sanidine
Marmatite (Fe bearing Sphalerite)	Sassolite (Sassoline)
Microcline	Scheelite
Millerite	Serpentine (Serpentinite) (Group Name)
Molybdenite	Siderite
Molybdenite	Sillimanite
Monazite-[Ce]	Silver
Monazite-[Ce]	Skutterudite
Mullite	Smithsonite
Muscovite	Sperrylite
Naumannite	Spessartine (Spessartite)

Alphabetical Index of Economic Minerals cont...

Commodity Index of Economic Minerals

Aluminum	Borax		
Bohmite (Boehmite)	Kernite		
Gibbsite	Ulexite		
Diaspore	**Cadmium**		
Bauxite (Group Name)	Greenockite		
Aluminum Silicate	**Calcium Carbonate**		
Kyanite	Calcite		
Sillimanite	**Cesium**		
Mullite	Lepidolite		
Andalusite	Pollucite		
Antimony	**China Clay**		
Boulangerite	Kaolinite		
Antimony	**Chromium**		
Stibnite	Chromite		
Berthierite	**Cobalt**		
Arsenic	Safflorite		
Realgar	Linnaeite		
Arsenopyrite	Cobaltite		
Orpiment	Carrollite		
Arsenic	Skutterudite		
Asbestos	**Copper**		
Grunerite	Tetrahedrite		
Amosite (fibrous form of ferro-gedrite) (Grunerite)	Copper		
Crocidolite (fibrous variety of Riebeckite)	Covellite		
Chrysotile (subgroup name of Serpintinites)	Malachite		
Serpentine (Serpentinite) (Group Name)	Cuprite		
Barium	Enargite		
Witherite	Tennantite		
Barytes	Stannite		
Baryte	Bournonite		
Beryllium	Azurite		
Beryl	Chrysocolla		
Bismuth	Bornite		
Bismuthinite	Chalcocite		
Bismutite	Chalcopyrite		
Bismuth	**Corundum**		
Borates	Corundum		
Colemanite	**Cryolite**		
Sassolite (Sassoline)	Cryolite		
Boracite			

Commodity Index of Economic Minerals cont...

Diamond	Indium
Diamond	Indium
Dolomite	**Iron**
Dolomite	Limonite
Epsom Salts	Pyrite
Epsomite	Magnetite
Feldspar	Hematite
Microcline	Goethite
Orthoclase	Pyrrhotite
Plagioclase	Siderite
Albite	**Lead**
Sanidine	Cerussite
Anorthite	Anglesite
Fluorspar	Jamesonite
Fluorite	Galena
Gallium	**Lithium**
Gallium	Amblygonite
Garnet	Spodumene
Almandine	Tourmaline (Elbaite)
Spessartine (Spessartite)	Lepidolite
Grossular	**Magnesite**
Uvarovite	Magnesite
Andradite	**Magnesium**
Pyrope	Brucite
Germanium	Dolomite
Argyrodite	Magnesite
Gold	Carnallite
Sylvanite	Manganite
Pyrite	Braunite
Goldamalgam	Pyrolusite
Gold	Bixbyite
Calaverite	Psilomelane
Graphite	Alabandite
Graphite	**Mercury**
Gypsum	Cinnabar
Gypsum	**Mica**
Hafnium	Biotite
Hafnium	Muscovite
Ilmenite	Phlogopite
Ilmenite	

Commodity Index of Economic Minerals cont...

Molybdenum	Rock Salt
Molybdenite	Halite
Wulfenite	**Rubidium**
Nickel	Lepidolite
Pentlandite	**Rutile**
Rammelsbergite	Rutile
Millerite	**Selenium**
Niccolite (Nickeline)	Berzelianite
Breithauptite	Naumannite
Niobium	Eucairite
Tantalite, see ferrotantalite and manganotantalite	Clausthalite
Columbite – Tantalite	**Serpentine**
Pyrochlore	Serpentine (Serpentinite) (Group Name)
Coltan (Columbite-Tantalite)	**Silicon**
Osmiridium	Quartz (alpha-Quartz)
Osmiridium	**Silver**
Phosphate	Galena
Apatite (Group Name)	Chlorargyrite (Cerargyrite)
Wavellite	Silver
Fluorapatite	Argentite
Phosphate Rock	Pyrargyrite
Platinum	Allargentum
Sperrylite	Proustite
Platinum	**Strontium**
Potash	Celestite
Carnallite	**Sulfur**
Alunite	Sulfur
Sylvite (Sylvine)	**Talc**
Niter	Talc (Steatite)
Quartz (Silica)	**Tantalum**
Quartz [alpha-Quartz]	Pyrochlore
Rare Earths	Ferrotantalite
Monazite-[Ce]	Manganotantalite
Bastnaesite-[Ce]	Tantalite, see ferrotantalite and manganotantalite
Bastnaesite-[Y]	**Tellurium**
Bastnaesite-[Ce]	Sylvanite
Bastnaesite-[La]	**Thorium**
Xenotime-[Y]	Monazite-[Ce]
Rhenium	Thorianite
Molybdenite	

Tin	
Cassiterite	
Tin	
Titanium	
Ulvospinel	
Ilmenite	
Rutile	
Titanite	
Tungsten	
Scheelite	
Wolframite (Wolfram)	
Uranium	
Carnotite	
Torbernite	
Uraninite (Pitchblende)	
Autunite	
Vanadium	
Vanadinite	
Patronite	
Roscoelite	
Carnotite	
Vermiculite	
Vermiculite	
Zinc	
Hemimorphite	
Franklinite	
Sphalerite	
Calamine (Hemimorphite)	
Marmatite (Fe bearing Sphalerite)	
Willemite	
Zincite	
Smithsonite	
Zirconium	
Zircon	

References

Bayliss, P., Berry, L.G., Mrose, M.E., Sabina, A.P., Smith, D.K. (eds), 1983, "Mineral Powder Diffraction File". JCPDS, ICDD, 1005p.

Craig, J.R., Vaughan, D,J., 1981, " Ore Microscopy and Ore Petrography", John Wiley and Sons, New York, 406p.

Dana, J.D., "Manual of Mineralogy".

Deer, W.A., Howie, R.A. and Zussman, J, 1980, "An Introduction to the Rock Forming Minerals". Longman, London. 528p.

Henry, N.F.M. (ed.), 1977, "Commision on Ore Microscopy:IMA/COM Quantitative Data File". Applied Mineralogy Group, Mineralogical Society, London.

JCPDS-International Center for Diffraction Data, 1995, " Powder Diffraction File. Alphabetical Index, Inorganic Phases". JCPDS, International Center for Diffraction Data.

Jenkins, R., 1996." Introduction to X-ray Powder Diffractometry ". Wiley, New York.

Kerr, P.F., 1977, "Optical Mineralogy", McGraw-Hill, 492p.

Le Maitre, R.W.(ed), 1989, "A Classification of Igneous Rocks and Glossary of Terms". Blackwell Scientific Publications, Oxford, UK. - Igneous plots and systematics.

Myashiro, A.,1973. "Metamorphism and Metamorphic Belts". Allen and Unwin, London,492p.

Mutschler, F.E., Rougon, D.J., Lavin, O.P., Hughes, R.D., 1981, "Petros - A Data Bank of Major Element Chemical Analyses of Igneous Rocks for Research and Teaching (Version 6.1)". NOAA-National Geophysical and Solar-Terrestrial Data Center.

Nichols, M.C., Nickel, E.H., 1991, "Mineral Reference Manual", Chapman and Hall, New York, 250p.

Wills, B.A., 1992, "Mineral Processing Technology". Pergamon, Oxford, 855p.

Web Sites and Software:

Athena Mineralogy http://un2sg1.unige.ch/www/athena/mineral/mineral.html - List of Mineral Names, Mineralogy

Geologynet.com http://www.geologynet.com/indexa.htm - Mineralogy

LR Ream Publishing http://www.mineralnews.com/ - Publishers of The Mineral Database (TMD)

Mineral Database - Materials Data Mineral Database, Formerly Aleph Interprises

Mineralogy Database http://webmineral.com/ - HTML Mineral Database

MinDat.org http://www.mindat.org/ - Mineralogy Database

Picture Credits:
Cover photo by the author

www.ingramcontent.com/pod-product-compliance
Lightning Source LLC
Chambersburg PA
CBHW081117170526
45165CB00008B/2470